Compendium of Plant Genomes

Series Editor

Chittaranjan Kole, Raja Ramanna Fellow, Government of India, ICAR-National Research Center on Plant Biotechnology, Pusa, New Delhi, India

Whole-genome sequencing is at the cutting edge of life sciences in the new millennium. Since the first genome sequencing of the model plant Arabidopsis thaliana in 2000, whole genomes of about 100 plant species have been sequenced and genome sequences of several other plants are in the pipeline. Research publications on these genome initiatives are scattered on dedicated web sites and in journals with all too brief descriptions. The individual volumes elucidate the background history of the national and international genome initiatives; public and private partners involved; strategies and genomic resources and tools utilized; enumeration on the sequences and their assembly; repetitive sequences; gene annotation and genome duplication. In addition, synteny with other sequences, comparison of gene families and most importantly potential of the genome sequence information for gene pool characterization and genetic improvement of crop plants are described.

Interested in editing a volume on a crop or model plant?
Please contact Prof. C. Kole, Series Editor, at ckoleorg@gmail.com

More information about this series at http://www.springer.com/series/11805

Fure-Chyi Chen • Shih-Wen Chin
Editors

The Orchid Genome

Springer

Editors
Fure-Chyi Chen
Department of Plant Industry
National Pingtung University
of Science and Technology
Pingtung County, Nei-pu Town, Taiwan

Shih-Wen Chin
Department of Plant Industry
National Pingtung University
of Science and Technology
Pingtung County, Nei-pu Town, Taiwan

ISSN 2199-4781 ISSN 2199-479X (electronic)
Compendium of Plant Genomes
ISBN 978-3-030-66825-9 ISBN 978-3-030-66826-6 (eBook)
https://doi.org/10.1007/978-3-030-66826-6

This Springer imprint is published by the registered company Springer Nature Switzerland AG
The registered company address is: Gewerbestrasse 11, 6330 Cham, Switzerland

This book series is dedicated to my wife Phullara and our children Sourav and Devleena

Chittaranjan Kole

Preface to the Series

Genome sequencing has emerged as the leading discipline in the plant sciences coinciding with the start of the new century. For much of the twentieth century, plant geneticists were only successful in delineating putative chromosomal location, function, and changes in genes indirectly through the use of a number of "markers" physically linked to them. These included visible or morphological, cytological, protein, and molecular or DNA markers. Among them, the first DNA marker, the RFLPs, introduced a revolutionary change in plant genetics and breeding in the mid-1980s, mainly because of their infinite number and thus potential to cover maximum chromosomal regions, phenotypic neutrality, absence of epistasis, and codominant nature. An array of other hybridization-based markers, PCR-based markers, and markers based on both facilitated construction of genetic linkage maps, mapping of genes controlling simply inherited traits, and even gene clusters (QTLs) controlling polygenic traits in a large number of model and crop plants. During this period, a number of new mapping populations beyond F2 were utilized and a number of computer programs were developed for map construction, mapping of genes, and for mapping of polygenic clusters or QTLs. Molecular markers were also used in the studies of evolution and phylogenetic relationship, genetic diversity, DNA fingerprinting, and map-based cloning. Markers tightly linked to the genes were used in crop improvement employing the so-called marker-assisted selection. These strategies of molecular genetic mapping and molecular breeding made a spectacular impact during the last one and a half decades of the twentieth century. But still they remained "indirect" approaches for elucidation and utilization of plant genomes since much of the chromosomes remained unknown and the complete chemical depiction of them was yet to be unraveled.

Physical mapping of genomes was the obvious consequence that facilitated the development of the "genomic resources" including BAC and YAC libraries to develop physical maps in some plant genomes. Subsequently, integrated genetic–physical maps were also developed in many plants. This led to the concept of structural genomics. Later on, emphasis was laid on EST and transcriptome analysis to decipher the function of the active gene sequences leading to another concept defined as functional genomics. The advent of techniques of bacteriophage gene and DNA sequencing in the 1970s was extended to facilitate sequencing of these genomic resources in the last decade of the twentieth century.

As expected, sequencing of chromosomal regions would have led to too much data to store, characterize, and utilize with the-then available computer software could handle. But the development of information technology made the life of biologists easier by leading to a swift and sweet marriage of biology and informatics, and a new subject was born—bioinformatics.

Thus, the evolution of the concepts, strategies, and tools of sequencing and bioinformatics reinforced the subject of genomics—structural and functional. Today, genome sequencing has traveled much beyond biology and involves biophysics, biochemistry, and bioinformatics!

Thanks to the efforts of both public and private agencies, genome sequencing strategies are evolving very fast, leading to cheaper, quicker, and automated techniques right from clone-by-clone and whole-genome shotgun approaches to a succession of second-generation sequencing methods. The development of software of different generations facilitated this genome sequencing. At the same time, newer concepts and strategies were emerging to handle sequencing of the complex genomes, particularly the polyploids.

It became a reality to chemically—and so directly—define plant genomes, popularly called whole-genome sequencing or simply genome sequencing.

The history of plant genome sequencing will always cite the sequencing of the genome of the model plant Arabidopsis thaliana in 2000 that was followed by sequencing the genome of the crop and model plant rice in 2002. Since then, the number of sequenced genomes of higher plants has been increasing exponentially, mainly due to the development of cheaper and quicker genomic techniques and, most importantly, the development of collaborative platforms such as national and international consortia involving partners from public and/or private agencies.

As I write this preface for the first volume of the new series "Compendium of Plant Genomes," a net search tells me that complete or nearly complete whole-genome sequencing of 45 crop plants, eight crop and model plants, eight model plants, 15 crop progenitors and relatives, and three basal plants is accomplished, the majority of which are in the public domain. This means that we nowadays know many of our model and crop plants chemically, i.e., directly, and we may depict them and utilize them precisely better than ever. Genome sequencing has covered all groups of crop plants. Hence, information on the precise depiction of plant genomes and the scope of their utilization are growing rapidly every day. However, the information is scattered in research articles and review papers in journals and dedicated Web pages of the consortia and databases. There is no compilation of plant genomes and the opportunity of using the information in sequence-assisted breeding or further genomic studies. This is the underlying rationale for starting this book series, with each volume dedicated to a particular plant.

Plant genome science has emerged as an important subject in academia, and the present compendium of plant genomes will be highly useful to both students and teaching faculties. Most importantly, research scientists involved in genomics research will have access to systematic deliberations on the plant genomes of their interest. Elucidation of plant genomes is of interest not only for the geneticists and breeders, but also for practitioners of an array of plant science disciplines, such as taxonomy, evolution, cytology, physiology, pathology, entomology, nematology, crop production, biochemistry, and obviously bioinformatics. It must be mentioned that information regarding each plant genome is ever-growing. The contents of the volumes of this compendium are, therefore, focusing on the basic aspects of the genomes and their utility. They include information on the academic and/or economic importance of the plants, description of their genomes from a molecular genetic and cytogenetic point of view, and the genomic resources developed. Detailed deliberations focus on the background history of the national and international genome initiatives, public and private partners involved, strategies and genomic resources and tools utilized, enumeration on the sequences and their assembly, repetitive sequences, gene annotation, and genome duplication. In addition, synteny with other sequences, comparison of gene families, and, most importantly, the potential of the genome sequence information for gene pool characterization through genotyping by sequencing (GBS) and genetic improvement of crop plants have been described. As expected, there is a lot of variation of these topics in the volumes based on the information available on the crop, model, or reference plants.

I must confess that as the series editor, it has been a daunting task for me to work on such a huge and broad knowledge base that spans so many diverse plant species. However, pioneering scientists with lifetime experience and expertise on the particular crops did excellent jobs editing the respective volumes. I myself have been a small science worker on plant genomes since the mid-1980s and that provided me the opportunity to personally know several stalwarts of plant genomics from all over the globe. Most, if not all, of the volume editors are my longtime friends and colleagues. It has been highly comfortable and enriching for me to work with them on this book series. To be honest, while working on this series I have been and will remain a student first, a science worker second, and a series editor last. And I must express my gratitude to the volume editors and the chapter authors for providing me the opportunity to work with them on this compendium.

I also wish to mention here my thanks and gratitude to the Springer staff, particularly Dr. Christina Eckey and Dr. Jutta Lindenborn for the earlier set of volumes and presently Ing. Zuzana Bernhart for all their timely help and support.

I always had to set aside additional hours to edit books beside my professional and personal commitments—hours I could and should have given to my wife, Phullara, and our kids, Sourav and Devleena. I must mention that they not only allowed me the freedom to take away those hours from them but also offered their support in the editing job itself. I am really not sure whether my dedication of this compendium to them will suffice to do justice to their sacrifices for the interest of science and the science community.

New Delhi, India Chittaranjan Kole

Preface

Orchidsrchids belong to the family Orchidaceae, creating approximately 10% of angiosperms with more than 25,000 species, comprises many unique physiological characteristic features, a complex genome with large size, spectacular flowers, specialized pollination, reproductive and ecological adaptations. Because of these significant remarkable features, orchids captured as most prevalent plant species for research in plant biology since Darwin published Fertilization of Orchids in 1862.

Orchids account for a great part of the worldwide floriculture trade both as cut flowers and as potted plants and are assessed to comprise around 10% of global fresh cut flower trade. The potted *Phalaenopsis* accounts for more than US$ 500 million in the global production and consumption markets, such as in Netherlands, Taiwan, United States, Japan, and other countries. A better understanding of the basic botanical characteristics, flower regulation, molecular cytogenetics, karyotypes and DNA content of important orchids will aid in the efficient development of new cultivars. In this genomic era, cracking genomes of orchids is a key step in understanding and cultivating new cultivars for commercial industry. In current days, advanced tools in genomics and proteomics such as next-generation sequencing technologies like Second-Generation Sequencing (SGS) and third-generation sequencing (TGS) technologies used in the assembly of the orchid genome resulted in the progress of an extensive variety of valuable resources for the genome of orchids. In the past few years, various research groups produced both genomic and transcriptomic data on orchid species such as *Phalaenopsis*, *Dendrobium*, *Orchis italica*, *Erycina pusilla*, and primitive orchid *Apostasia*. These deciphering orchid genomes show the significant role of genes involved in heterozygosity, specific paleopolyploidy, the evolution of CAM photosynthesis, diversification in MADS-box genes in relation to flower development, physiology of tropical and temperate orchids, and climate change and its impact on orchid productivity. Recently, the information about the composition, expression, and function of various microRNAs and simple sequence repeats were identified in orchids, the information of their involvement in all aspects of plant growth and development will aid functional genomics studies. The expression levels of specifically enriched transcription factors and signaling components in interior ovary tissues provide their potential role in gametophyte development, epigenetic reprogramming, and hormone regulation during fertilization and establishment of embryo development.

The volume on "The Orchid Genome" will provide a broad scope for the study of orchid biology and will organize information on genome complexity and evolution, transcriptome analysis, miRNome, simple sequence repeats, genome relationships, molecular markers and karyotype analysis, breeding methods, flower and embryo development. Further, it provides opportunities and challenges for genetic improvement of the novel orchids to meet the changing market demands.

Pingtung County, Taiwan

Fure-Chyi Chen, PhD
Shih-Wen Chin, PhD

Contents

Contributors

Serena Aceto Department of Biology, University of Naples Federico II, Naples, Italy

Teresita D. Amore Department of Tropical Plant and Soil Sciences, University of Hawaii at Manoa, Honolulu, Hawaii, USA

Pablo Bolaños-Villegas Fabio Baudrit Agricultural Research Station, University of Costa Rica, Alajuela, Costa Rica;
Jardín Botánico Lankester, Universidad de Costa Rica, Cartago, Costa Rica

Chen Chang Department of Horticulture, National Chung Hsin University, Taichung, Taiwan

Ya-Ting Chao Agricultural Biotechnology Research Center, Academia Sinica, Taipei, Taiwan

Fure-Chyi Chen Department of Plant Industry, National Pingtung University of Science and Technology, Pingtung, Taiwan;
General Research Service Center, National Pingtung University of Science and Technology, Pingtung, Taiwan

Wan-Chieh Chen Agricultural Biotechnology Research Center, Academia Sinica, Taipei, Taiwan

Yu-Chung Chiang National Sun Yat-Sen University, Kaohsiung, Taiwan

Shih-Wen Chin Department of Plant Industry, National Pingtung University of Science and Technology, Pingtung, Taiwan

Mei-Chu Chung Institute of Plant and Microbial Biology, Academia Sinica, Taipei, Taiwan, ROC

Hsiu-Yin Ho Agricultural Biotechnology Research Center, Academia Sinica, Taipei, Taiwan

Jian-Zhi Huang Department of Plant Industry, National Pingtung University of Science and Technology, Pingtung, Taiwan;
Department of Agricultural Chemistry, College of Bioresources and Agriculture, National Taiwan University, Taipei, Taiwan

Xiao-Lei Jin National Sun Yat-Sen University, Kaohsiung, Taiwan

Ya-Zhu Ko National Sun Yat-Sen University, Kaohsiung, Taiwan

Yung-I Lee Department of Biology, National Museum of Natural Science, Taichung, Taiwan, ROC;
Department of Life Sciences, National Chung Hsing University, Taichung, Taiwan, ROC

Setapong Lekawatana Department of Agricultural Extension, Ministry of Agriculture and Cooperatives, Bangkok, Thailand

I-Chun Pan Department of Horticulture, National Chung Hsing University, Taichung, Taiwan

Ming-Che Shih Agricultural Biotechnology Research Center, Academia Sinica, Taipei, Taiwan

Chi-Chu Tsai National Sun Yat-Sen University, Kaohsiung, Taiwan

Chin-Yi Tsao Section of Biotechnology, Taiwan Agricultural Research Institute, Taichung, Taiwan

David Monge Vega Florali Business Consultancy B.V., Rotterdam, The Netherlands

Kotapati Kasi Viswanath Department of Plant Industry, National Pingtung University of Science and Technology, Pingtung, Taiwan

Yin-Tung Wang Department of Horticultural Sciences, Texas A&M University System, Texas Agricultural Research and Extension Center, Weslaco, TX, USA

Shih-Chang Yuan Department of Plant Industry, National Pingtung University of Science and Technology, Pingtung, Taiwan

Abbreviation

ABA	Abscisic acid
ABCDE	Model, five major classes of homeotic selector genes A, B, C, D and E controlling flower development
AFLP	Amplified fragment length polymorphism
AG	Agamous
AP1	Apetala1
BA	6-benzylaminopurine
BESs	Bacterial artificial chromosome end sequences
BiFC	Bimolecular fluorescence complementation
bZIP	Basic leucine zipper
C3	C3 photosynthetic plants
CAM	Crassulacean acid metabolism
CH	Constitutive heterochromatin
CKs	Cytokinins
CKX	Cytokinin oxidase/dehydrogenase
CO	Cross-over
COL	Constans-like
CRISPR	Clustered Regularly Interspaced Short Palindromic Repeats
DAPI	4', 6-diamidino-2-phenylindole
DEF	Deficiens
DEGs	Differentially expressed genes
Denphal	*Dendrobium phalaenopsis* hybrids
DNP	Day Neutral Plants
DRIF	DIV and RAD Interacting Factor
ELF4	Early Flowering4
ESTs	Expressed sequence tags
evo-devo	evolutionary developmental
FBP	Floral Binding Protein
FDA	Fluorescein diacetate
FISH	Fluorescent in situ hybridization
FT	Flowering Locus T
GA	Gibberellin
GLO	Globosa
GO	Gene Ontology
ISSR	Inter-simple sequence repeat
ITS	Internal transcribed spacer

JA	Jasmonic acid
KEGG	Kyoto Encyclopedia of Genes and Genomes
LDP	Long Day Plants
LFY	LEAFY
lncRNAs	Long noncoding RNAs
LTRs	Long terminal repeats
MADS	**M**CM1 from *Saccharomyces cerevisiae*, **A**GAMOUS from *Arabidopsis thaliana*, **D**EFICIENS from *Antirrhinum majus*, **S**RF from the human *Homo sapiens*
MCM1	Minichromosome Maintenance 1
MFT	Mother of FT and TFL1
MH	Maleic hydrazide
Myr	Million years
NCBI	National Center for Biotechnology Information
NGS	Next-generation sequencing
NHEJ	Non-homologous end joining
NORs	Nucleolar-organizing regions
OAA	Oxaloacetate
PEP	Phosphoenolpyruvate
PEPC	Phosphoenolpyruvate carboxylase
PGM	Personal genome machine
PI	Pistillata
PLBs	Protocorm-like bodies
qPCR	Real-time polymerase chain reaction, Quantitative PCR
RHS	Royal Horticultural Society
RuBisCO	Ribulose-1,5-bisphosphate carboxylase/oxygenase
SA	Salicylic acid
SAM	Shoot apical meristem
SDP	Short Day Plants
SEP	Sepallata
SL	Strigolactone
smRNA	small modulatory RNA
SQUA	Squamosa
SRF	Serum Response Factor
sRNA	small RNA
SOC1	Suppressor of Overexpression of CO1
SOLiD	Sequencing by Oligo Ligation Detection
SVP	Short Vegetative Phase
TCP	*Teosinte branched1* (*tb1*) from *Zea mays*, *CYCLOIDEA* (*CYC*) from *Antirrhinum majus*, *PROLIFERATING CELL FACTORS 1* and *2* (*PCF1* and *PCF2*) from *Oryza sativa*
TCV	Total chromosome volume
TDZ	Thidiazuron
TEs	Transposable elements
TFs	Transcription factors
TIBA	2,3,5-triiodobenzoic acid
VIGS	Virus-induced gene silencing

1 The Global Orchid Market

Shih-Chang Yuan, Setapong Lekawatana, Teresita D. Amore, Fure-Chyi Chen, Shih-Wen Chin, David Monge Vega, and Yin-Tung Wang

Abstract

The global orchid production has been active in tropical, subtropical countries, many with automated and efficiently managed large greenhouses. The most important exported potted orchid in the trade is *Phalaenopsis* with tremendous breeding and micropropagation technology achievements in several countries, such as Belgium, the Netherlands, Taiwan, and Thailand. Another reason for its popularity is its easiness of controlled flower induction for scheduled and year-around production. Cut orchids such as *Cymbidium*, *Dendrobium*, *Oncidium*, and *Vanda* are important cut flowers worldwide. Now orchids are quite popular in the global market, partly due to their long shelf life, diverse colors, and other desirable traits pursued by consumers, depending on the location or culture. European Union (EU) and the United States are the most important countries or areas for marketing of assortments of orchid products. New markets for orchids are increasing annually. Since the beginning of 2020, due to the severe coronavirus pandemic and disruption of global economy, orchid marketing has been shifting to an increased share of digitalization or E-commerce, co-existing with the traditional flower stores and supermarkets. In the future, more orchid products or novel new orchid hybrids being developed by non-conventional technologies, such as biotech-oriented efficient breeding, may help offer new opportunities for orchid production.

1.1 Introduction

Despite the many cultures in the world, the orchid has been a common beloved flower. Tropical orchids hunted from the New World

S.-C. Yuan · F.-C. Chen · S.-W. Chin
Department of Plant Industry, National Pingtung University of Science & Technology, Pingtung 91201, Taiwan

S. Lekawatana
Department of Agricultural Extension, Ministry of Agriculture and Cooperatives, Bangkok 10900, Thailand

T. D. Amore
Department of Tropical Plant and Soil Sciences, University of Hawaii at Manoa, Honolulu, Hawaii 96822, USA

D. M. Vega
Florali Business Consultancy B.V., Rotterdam, The Netherlands

Y.-T. Wang (✉)
Department of Horticultural Sciences, Texas A&M University System, Texas Agricultural Research and Extension Center, 2415 East Highway 83, Weslaco, TX 78596, USA
e-mail: yintung.wang@gmail.com

F.-C. Chen and S.-W. Chin (eds.), *The Orchid Genome*, Compendium of Plant Genomes,
https://doi.org/10.1007/978-3-030-66826-6_1

were treated as the symbol of identity and wealth in Europe aristocracy. In ancient Chinese history, rulers loved orchids. Owning orchids was also deemed as nobleman and camaraderie. Orchids have been cultivated for both hobby and commercial uses. Both potted and cut orchids are traded in the flower markets globally. About 60 years ago, owing to the development of aseptic culture and plastic greenhouse technologies, market demand for orchids began to evolve from nobility to popularization. Around two decades ago, orchids started their introduction in the Dutch and US markets. Back then, they were considered a luxury and not a simple flower or plant. During the years ensued, this uniqueness changed from being a product only found in the boutique flower shop into a common product at almost every retailer and flower shops.

If we compare the market figures from 20 years ago to what they are now, we see a huge increase in production and consumption of orchids as well. And, as a result of the lower price and the overwhelming presence of the orchids in the marketplace, the status of orchids has changed from a luxury good to a common consumer product worldwide. Orchids aren't special anymore.

1.2 Global Import of Orchid Cut Flowers

Global import of all cut flowers was an $8,490 million business in 2019 with a 6.1% decline from the previous year and had a 3.0% average annual growth for the five-year period of 2015–2019, regardless of the global economic crisis. Import of cut orchids valued at $214 million in 2019, a 2.5% share of all cut flower import. The cut orchid import value declined by 13.5% from 2018 to 2019 and at an averaged 1.4% annual decline over the past 5 years (Fig. 1.1). Japan is the world largest importer of cut orchids for many years. In 2019, Japan imported $62.4 million worth of cut orchids, at 29.2% of the global import value with no growth from the previous year. The second largest cut orchid importer was the United States at $22.9 million, 10.7% of the global import value with a 21.3% decline from that of 2018. Other major cut orchid importers include (in the order of value) Italy, China, Vietnam, the United Kingdom, France, Germany, Singapore, and Australia (Fig. 1.2).

In the Japanese market, Taiwan was the largest exporter of cut orchids at $28.9 million, accounting for 46.3% of all Japan import with a 0.4% growth

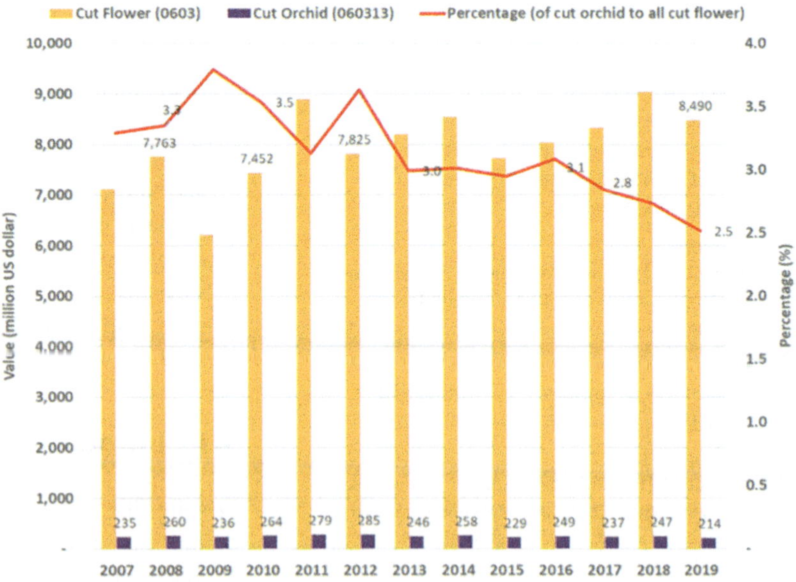

Fig. 1.1 The global import value of cut flowers and cut orchids and the percentage of cut orchids to all cut flowers from 2007 to 2019. *Data* International Trade Center, 2020

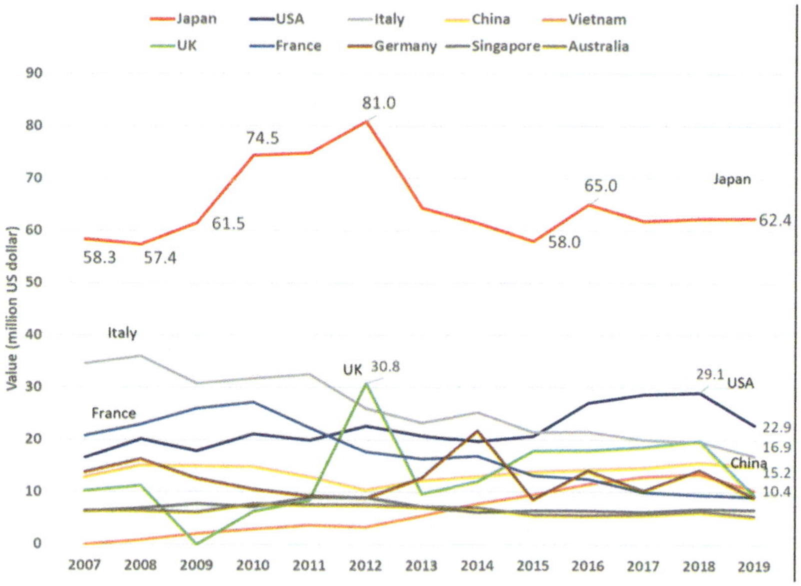

Fig. 1.2 Global cut orchid import value of major importers from 2007 to 2018. *Data* International Trade Center, 2020

from 2018 to 2019. Thailand was the second largest country exporting cut orchids to Japan in 2019 with a value of $20.0 million, 32.1% of all Japanese import. Other major exporters to the Japanese market in 2019 were (in the order of value) Vietnam, New Zealand, Malaysia, China, South Korea, Singapore, and Australia (Fig. 1.3).

In the US market, Thailand was the dominating exporter of cut orchids at $13.0 million, holding a 57.0% share of all US import with a 29.7% decline from 2018 to 2019. For the past 5 years, the average annual growth was 2.8%. The second largest exporter was the Netherlands at $6.7 million or 29.5% of all US imports with a 12.3% decline from 2018 to 2019. Other important exporters to the US market were (in the order of value) New Zealand, Vietnam, Malaysia, and Taiwan (Fig. 1.4).

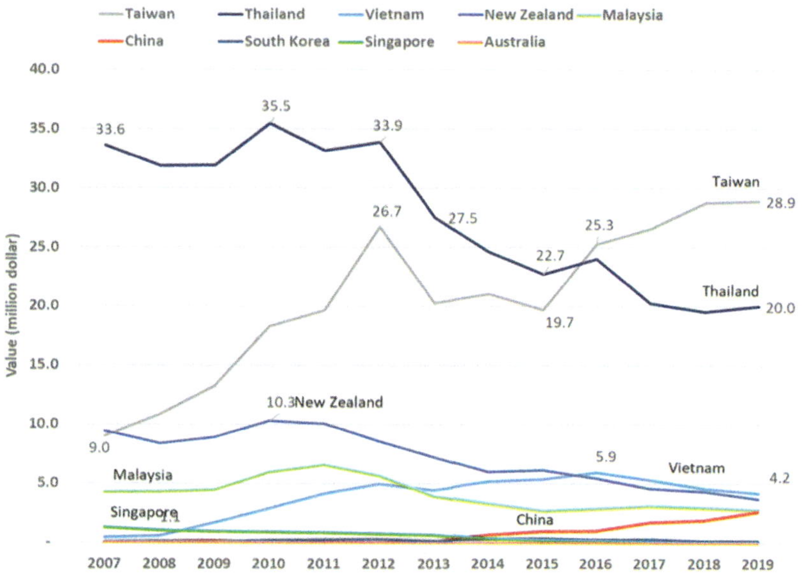

Fig. 1.3 Japan cut orchid import value from major exporting countries from 2007 to 2019. *Data* International Trade Center, 2020

Fig. 1.4 United States cut orchid import value from major exporting countries from 2007 to 2019. *Data* International Trade Center, 2020

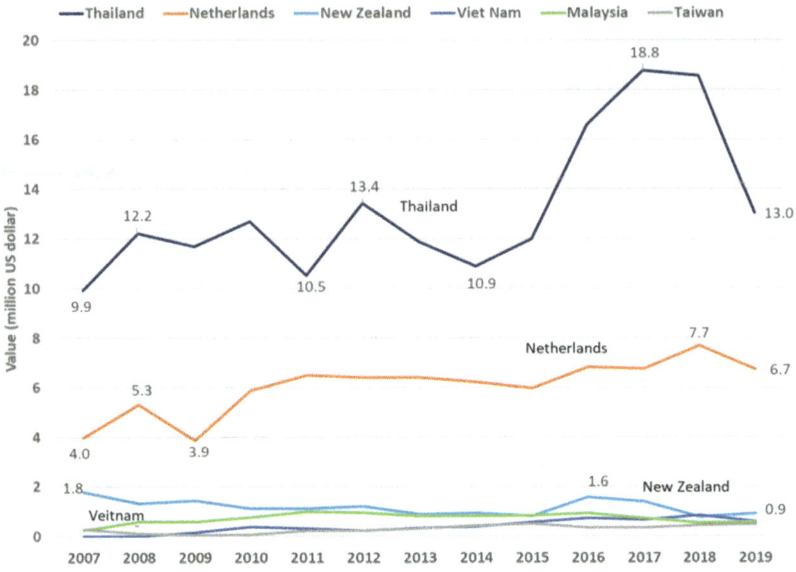

1.3 Global Export of Orchid Cut Flowers

The type of cut orchids being exported depends on the producers' location and preferences. The Netherlands and New Zealand are well known for their *Cymbidium*; the Netherlands and Taiwan for *Phalaenopsis*; Malaysia, Singapore, and Thailand for *Dendrobium* and *Mokara*; and Taiwan and Thailand for *Oncidium* and *Vanda*.

The Netherlands was the world's largest cut orchid exporter with an export value of $83.8 million in 2019, a 37.6% share of the total global export value and a 10.3% decline over 2018. Thailand was the second largest cut orchid exporter valued at $69.9 million with a 31.3% market share and a 2.5% decline over 2018. Taiwan was the third largest cut orchid exporter at $36.5 million with a 16.3% market share and a 27.6% growth over 2018. Other exporters (in the order of value) were Singapore, New Zealand,

Fig. 1.5 Values of cut orchids from the top 10 exporting countries from 2007 to 2019. *Data* International Trade Center, 2020

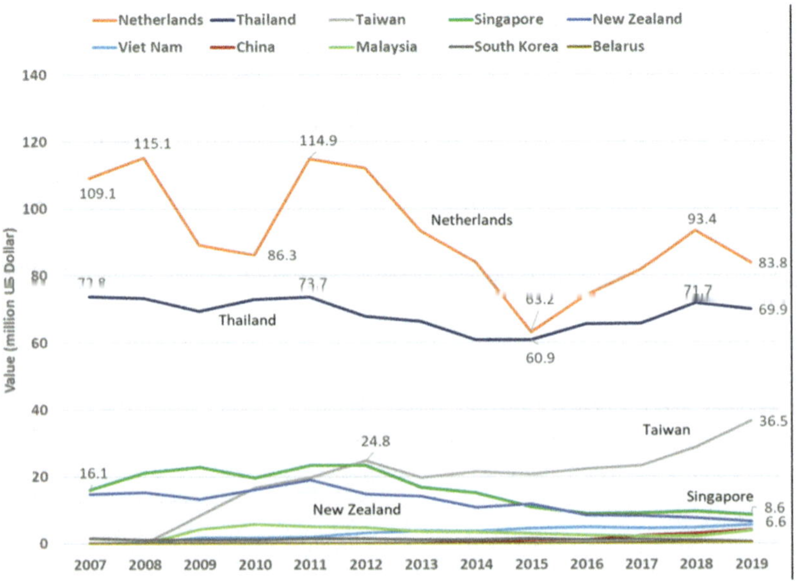

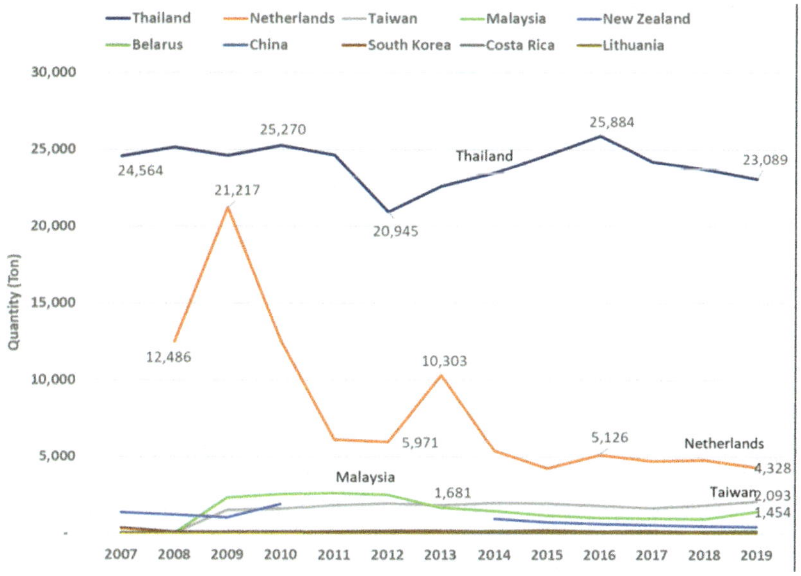

Fig. 1.6 The quantity of cut orchids from the top 10 exporting countries from 2007 to 2019. *Data* International Trade Center, 2020

Vietnam, China, Malaysia, South Korea, and Belarus (Fig. 1.5). Thailand exported the largest quantity of cut orchids worldwide at 23,089 tons, while the Netherlands exported 4,328 tons, and Taiwan exported 2,093 tons (Fig. 1.6).

1.4 Orchid Export from Taiwan

The floral products exported from Taiwan had a combined value of $199 million in 2017 (Taiwan Customs Administration 2019). The exported

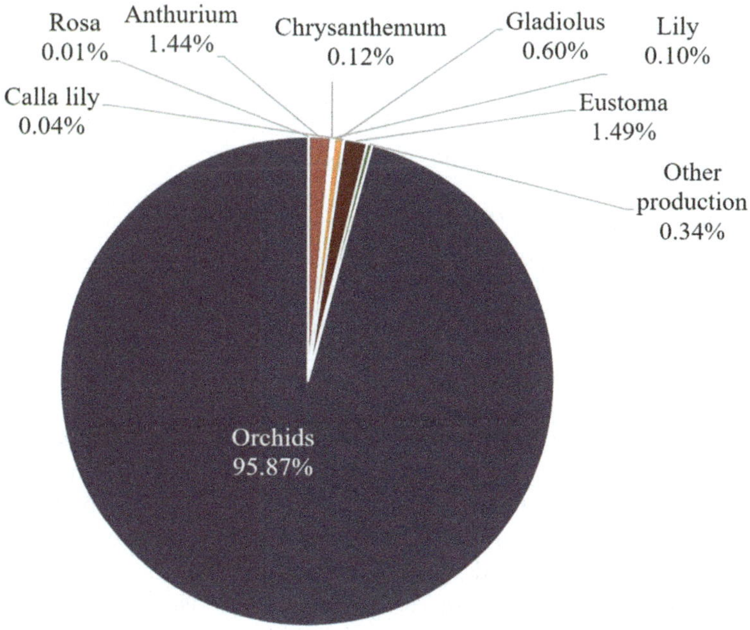

Fig. 1.7 Flowers being exported from Taiwan in 2018 and their shares. *Data* Customs Administration, Ministry of Finance, Taiwan (2019)

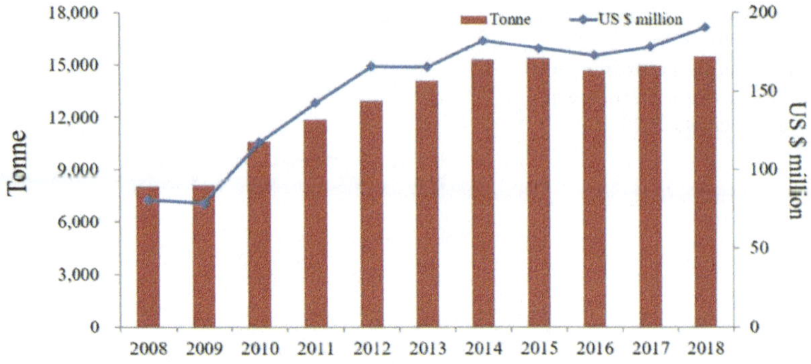

Fig. 1.8 Value of the exported orchids from Taiwan from 2008 to 2018. *Data* Council of Agriculture, Taiwan (2018)

Table 1.1 The top ten importing countries for all orchids from Taiwan in each of 2017, 2018, and 2019. *Data* Customs Administration, Ministry of Finance, Taiwan (2019)

Rank	Country	Export value[z]	Rank	Country	Export value	Growth rate (%)[y]	Rank	Country	Export value	Growth rate (%)
		2017			2018	2017–2018			2019	2018–2019
1	United States	57.74	1	United States	57.52	−0.38	1	United States	44.18	−23.19
2	Japan	50.67	2	Japan	52.41	3.43	2	Japan	43.49	−17.02
3	Netherlands	15.16	3	Vietnam	16.62	38.71	3	Vietnam	13.91	−16.31
4	Vietnam	11.95	4	Netherlands	14.99	−1.16	4	Netherlands	10.33	−31.08
5	South Korea	8.29	5	South Korea	9.97	20.21	5	Australia	6.69	−23.34
6	Australia	7.91	6	Australia	8.73	10.94	6	South Korea	6.25	−37.32
7	Canada	3.91	7	Canada	5.30	35.25	7	Canada	5.20	−1.78
8	Brazil	3.5	8	Brazil	4.85	38.51	8	Brazil	2.93	−39.65
9	United Kingdom	2.61	9	United Kingdom	2.49	−4.58	9	Singapore	1.66	−28.75
10	Singapore	2.10	10	Singapore	2.33	11.03	10	United Kingdom	1.60	−35.83
Total		117.68	Total		190.50	7.21	Total		148.11	−22.25

[z]$ million
[y](Current export value—past export value)/past export value × 100%

floricultural crops include orchids, eustoma, anthurium, gladiolus, roses, lily, calla lily, chrysanthemum, and seeds (designated as "other production") (Fig. 1.7). Orchids, valued at $191 million, were the predominantly exported floriculture items of Taiwan in 2018, accounting for 96% of the total exported value, a 7.2% growth over 2016 (Figs. 1.7 and 1.8; Table 1.1). In both 2018 and 2019, 30% of their exported orchids in value were shipped to the United States (Table 1.1). Taiwan also exported 35–38% of orchids in value to countries such as Brazil, Canada, and Vietnam in recent years. Their orchid export value in 2019 declined by 22% as compared to that of 2018.

Cattleya, *Cymbidium*, *Dendrobium*, *Oncidium*, *Paphiopedilum*, *Phalaenopsis*, certain other orchid genera, and flasked orchid plantlets are all

Table 1.2 The top ten importing countries for *Phalaenopsis* from Taiwan for each of 2017, 2018, and 2019. *Data* Customs Administration, Ministry of Finance, Taiwan (2019)

Rank	Country	Export value[z] 2017	Rank	Country	Export value[z] 2018	Growth rate (%)[y] 2017–2018	Rank	Country	Export value[z] 2019	Growth rate (%) 2018–2019
1	United States	56.81	1	United States	56.65	−0.38	1	United States	43.28	−23.60
2	Japan	37.08	2	Japan	36.06	−2.74	2	Japan	28.16	−21.91
3	Vietnam	11.09	3	Vietnam	15.47	39.50	3	Vietnam	13.48	−12.82
4	Australia	7.70	4	Australia	8.32	7.97	4	Australia	6.52	−21.57
5	Netherlands	3.90	5	Canada	5.12	37.05	5	Canada	5.09	−0.66
6	Canada	3.74	6	Brazil	3.85	32.66	6	Brazil	2.37	−38.51
7	Brazil	2.90	7	Netherlands	3.04	−22.10	7	Netherlands	1.85	−39.10
8	United Kingdom	2.61	8	United Kingdom	2.50	−4.58	8	United Kingdom	1.60	−35.87
9	Singapore	1.80	9	Singapore	1.99	10.47	9	Singapore	1.47	−26.13
10	South Korea	1.79	10	Other Africa	1.83	9.60	10	South Korea	1.36	−17.86
Total		139.87	Total		145.66	4.14	Total		113.65	−21.97

[z]$ million
[y](Current export value—past export value)/past export value × 100%

exported from Taiwan (Fig. 1.9), with a huge quantity of *Phalaenopsis* plants in pots being exported due in part to the conducive environmental conditions, including warm temperature and high light intensity which accelerate vegetative growth of young plants to mature

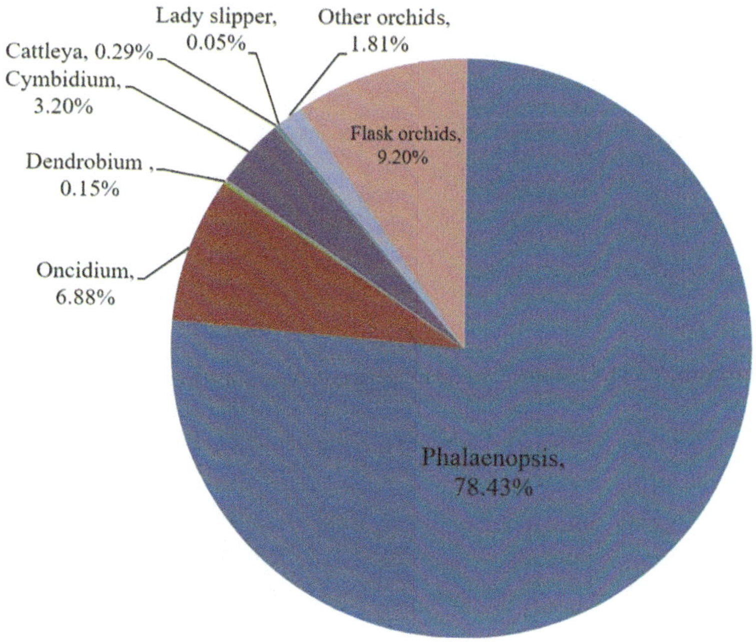

Fig. 1.9 The percentage of various orchids exported from Taiwan in 2018. *Data* Ministry of Finance, Taiwan (2019)

Table 1.3 The top ten importing countries for *Oncidium* from Taiwan for each of 2017, 2018, and 2019. *Data* Customs Administration, Ministry of Finance, Taiwan (2019)

Rank	Country	Export value[z]	Rank	Country	Export value	Growth rate (%)[y]	Rank	Country	Export value	Growth rate (%)
		2017			2018	2017–2018			2019	2018–2019
1	Japan	11.47	1	Japan	14.05	22.51	1	Japan	14.10	0.41
2	Singapore	0.22	2	Vietnam	0.25	165.56	2	Hong Kong	0.22	62.22
3	United States	0.16	3	Singapore	0.24	7.73	3	United States	0.21	−0.48
4	Hong Kong	0.09	4	United States	0.21	35.48	4	Singapore	0.15	−37.02
5	Vietnam	0.09	5	Australia	0.18	4739.33	5	United Arab Emirates	0.07	2200.00
6	South Africa	0.07	6	Hong Kong	0.14	46.68	6	South Africa	0.06	−38.14
7	China	0.04	7	South Africa	0.10	34.30	7	China	0.06	−15.71
8	South Korea	0.02	8	China	0.07	76.13	8	Australia	0.06	−66.86
9	Canada	0.02	9	South Korea	0.02	12.99	9	Macao	0.05	840.00
10	Macao	0.01	10	Chile	0.01	100.00	10	Malaysia	0.03	383.33
Total		12.26	Total		15.34	25.07	Total		15.12	−1.43

[z]$ million
[y](Current export value—past export value)/past export value × 100%

Table 1.4 The top ten importing countries for *Cymbidium* from Taiwan for each of 2017, 2018, and 2019. *Data* Customs Administration, Ministry of Finance, Taiwan (2019)

Rank	Country	Export value[z]	Rank	Country	Export value	Growth rate (%)[y]	Rank	Country	Export value	Growth rate (%)
		2017			2018	2017–2018			2019	2018–2019
1	South Korea	5.510	1	South Korea	7.243	31.45	1	South Korea	4.034	−44.30
2	China	0.164	2	China	0.142	−13.29	2	China	0.052	−63.38
3	United States	0.005	3	United States	0.005	−3.57	3	Iraq	0.010	100.00
4	Singapore	0.004	4	Russia	0.003	282.00	4	United States	0.006	20.00
5	Germany	0.002	5	Dominican Republic	0.002	100.00	5	Malaysia	0.003	200.00
6	Japan	0.002	6	Singapore	0.001	−77.73	6	Switzerland	0.002	100.00
7	Canada	0.001	7	Hong Kong	0.001	100.00	7	Russia	0.001	−66.67
8	France	0.001	8	Canada	0.001	−44.53	8	–	–	–
9	Spain	0.001	9	Malaysia	0.001	0.00	9	–	–	–
10	Malaysia	0.001	10	Guam	0.001	100.00	10	–	–	–
Total		5.696	Total		7.401	29.94	Total		4.108	−44.49

[z]$ million
[y](Current export value—past export value)/past export value × 100%

Table 1.5 The top ten importing countries for *Dendrobium* from Taiwan for each of 2017, 2018, and 2019. *Data* Customs Administration, Ministry of Finance, Taiwan (2019)

Rank	Country	Export value[z]	Rank	Country	Export value	Growth rate (%)[y]	Rank	Country	Export value	Growth rate (%)
		2017			2018	2017–2018			2019	2018–2019
1	Vietnam	0.185	1	Vietnam	0.384	107.89	1	Vietnam	0.133	−65.36
2	Singapore	0.015	2	Singapore	0.062	320.10	2	United States	0.018	80.00
3	Japan	0.010	3	South Korea	0.048	429.54	3	Dominican Republic	0.017	750.00
4	South Korea	0.009	4	United States	0.010	15.05	4	South Korea	0.015	−68.75
5	United States	0.009	5	Australia	0.005	3.94	5	China	0.004	100.00
6	Canada	0.007	6	Hong Kong	0.005	100.00	6	Brunei	0.004	100.00
7	Netherlands	0.007	7	South Africa	0.005	100.00	7	Russia	0.003	−25.00
8	Hong Kong	0.005	8	China	0.005	100.00	8	India	0.003	200.00
9	Russia	0.003	9	South Korea	0.005	−32.34	9	Japan	0.002	100.00
10	Switzerland	0.003	10	Chile	0.005	100.00	10	Malaysia	0.002	0.00
Total	0.269		Total	0.562		108.66	Total	0.211		−62.46

[z]$ million
[y](Current export value—past export value)/past export value × 100%

quickly for flower forcing. Valued at $146 million, *Phalaenopsis* constituted 76.4% of all exported orchids in 2018, followed by *Oncidium* at 6.9%, and *Cymbidium* at 3.2% (Table 1.2). The primary importing countries for *Phalaenopsis* orchid were the United States with a 38% share, Japan (24%), and Vietnam (11%). *Phalaenopsis* orchids from Taiwan have a stable market share in the United States, mainly attributed to the pre-certified greenhouses for exporting orchids in sphagnum moss as an approved substrate (see the Importation of Orchids in Growing Media from Taiwan, https://www.federalregister.gov/documents/2018/01/30/2018-01737/importation-of-orchids-in-growing-media-from-taiwan).

Oncidium orchid was the second largest exported item, valued at $12–$15 million in 2017–2019 (Fig. 1.9, Table 1.3). The main market was Japan, with 96–98% being cut flowers. The export of cut *Oncidium* flowers has been growing. Despite the United States approved the importation of *Oncidium* plants in growing media from Taiwan in 2016, the export value of pot *Oncidium* to the United States has remained relatively low.

Cymbidium orchids ranked third in export value and were predominantly exported to South Korea (96–97%) and valued at $5–$7.5 million (Table 1.4). The exported *Cymbidium* included two types, *C. hybridum* Hort. and other cymbidiums, with a market share of 0.2% and 99.8%, respectively. In South Korea, *Cymbidium* potted plants are used for various purposes, such as ceremonial occasions, gifts, as well as decorations. During the early years of exporting to South Korea, *Cymbidium* orchids were shipped bare-rooted by air. However, survival rates were low and costs were high. Since then, the South Korean government has permitted import of potted *Cymbidium* from Taiwan in growing medium, thus reducing the mortality and increasing the market value.

Table 1.6 The top ten exported countries for *Cattleya* from Taiwan for each of 2017, 2018, and 2019. *Data* Customs Administration, Ministry of Finance, Taiwan (2019)

Rank	Country	Export value[z]	Rank	Country	Export value	Growth rate (%)[y]	Rank	Country	Export value	Growth rate (%)
		2017			2018	2017–2018			2019	2018–2019
1	Vietnam	0.288	1	Vietnam	0.153	−46.89	1	Vietnam	0.082	−46.51
2	Dominica	0.039	2	South Korea	0.046	2327.15	2	United States	0.048	108.70
3	United States	0.035	3	Japan	0.046	33.20	3	South Korea	0.035	−23.91
4	Japan	0.035	4	Other Africa	0.044	100.00	4	Singapore	0.028	115.38
5	Singapore	0.016	5	Russia	0.028	400.00	5	Dominican Republic	0.025	−7.41
6	Germany	0.013	6	Dominican Republic	0.027	−31.45	6	Japan	0.022	−52.17
7	China	0.012	7	Australia	0.026	100.00	7	Netherlands	0.021	250.00
8	Ukraine	0.010	8	United States	0.023	−33.52	8	Germany	0.014	55.56
9	Netherlands	0.009	9	China	0.017	39.88	9	Russia	0.011	−60.71
10	Russia	0.007	10	Canada	0.015	232.93	10	Ecuador	0.008	300.00
Total	0.512		Total	0.514		0.32	Total	0.326		−36.58

[z]$ million
[y](Current export value—past export value)/past export value × 100%

Table 1.7 The top ten importing countries for *Paphiopedilum* from Taiwan for each of 2017, 2018, and 2019. *Data* Customs Administration, Ministry of Finance, Taiwan (2019)

Rank	Country	Export value[z]	Rank	Country	Export value	Growth rate (%)[y]	Rank	Country	Export value	Growth rate (%)
		2017			2018	2017–2018			2019	2018–2019
1	United States	0.020	1	United States	0.018	−9.53	1	Switzerland	0.022	144.44
2	Japan	0.014	2	South Korea	0.016	21.37	2	United States	0.012	−33.33
3	South Korea	0.013	3	Japan	0.011	−19.29	3	Canada	0.008	100.00
4	Canada	0.013	4	Switzerland	0.009	61.44	4	Japan	0.005	−54.55
5	Australia	0.008	5	Germany	0.005	129.10	5	South Korea	0.002	−87.50
6	Switzerland	0.006	6	Vietnam	0.003	100.00	6	Netherlands	0.00[illegible]	100.0[illegible]
7	Ecuador	0.005	7	Singapore	0.002	−44.24	7	Germany	0.002	−60.00
8	Singapore	0.004	8	Hong Kong	0.002	71.33	8	Ecuador	0.001	100.00
9	India	0.003	9	India	0.002	−28.99	9	New Caledonia	0.001	100.00
10	Russia	0.003	10	Ukraine	0.002	100.00	10	–	–	–
Total	0.097		Total	0.073		−24.51	Total	0.055		−24.66

[z]$ million
[y](Current export value—past export value)/past export value × 100%

Table 1.8 The top ten importing countries for other orchid genera from Taiwan for each of 2017, 2018, and 2019. *Data* Customs Administration, Ministry of Finance, Taiwan (2019)

Rank	Country	Export value[z]	Rank	Country	Export value	Growth rate (%)[y]	Rank	Country	Export value	Growth rate (%)
		2017			2018	2017–2018			2019	2018–2019
1	Japan	2.02	1	Japan	2.20	9.33	1	Japan	1.162	−47.28
2	China	0.18	2	China	0.16	−7.50	2	Netherlands	0.086	11.69
3	South Korea	0.14	3	Netherlands	0.08	34.80	3	South Korea	0.063	23.53
4	Netherlands	0.06	4	South Korea	0.05	−64.41	4	United States	0.048	0.00
5	Singapore	0.04	5	United States	0.05	198.36	5	China	0.045	−72.39
6	Germany	0.02	6	Dominican Republic	0.04	111.91	6	Russia	0.012	−36.84
7	Dominican Republic	0.02	7	Brazil	0.04	100.00	7	Singapore	0.011	−47.62
8	Russia	0.02	8	Singapore	0.02	−41.37	8	Germany	0.009	−18.18
9	United States	0.02	9	Russia	0.02	14.41	9	New Caledonia	0.005	66.67
10	Vietnam	0.01	10	Switzerland	0.01	40.61	10	Poland	0.004	100.00
Total	2.57		Total	2.73		6.22	Total	1.46		−46.54

[z]$ million

[y](Current export value—past export value)/past export value × 100%

Table 1.9 The top ten importing countries for flasked orchid plantlets from Taiwan for each of 2017, 2018, and 2019. *Data* Customs Administration, Ministry of Finance, Taiwan (2019)

Rank	Country	Export value[z]	Rank	Country	Export value	Growth rate (%)[y]	Rank	Country	Export value	Growth rate (%)
		2017			2018	2017–2018			2019	2018–2019
1	Netherlands	11.18	1	Netherlands	11.86	6.00	1	Netherlands	8.36	−29.46
2	Belgium	1.03	2	Belgium	1.35	31.48	2	Belgium	1.25	−7.04
3	Germany	0.86	3	Brazil	0.96	60.97	3	South Korea	0.73	−18.01
4	South Korea	0.81	4	South Korea	0.89	9.83	4	United States	0.56	0.18
5	United States	0.70	5	Germany	0.71	−16.90	5	Brazil	0.56	−41.94
6	Brazil	0.58	6	United States	0.56	−19.89	6	Germany	0.49	−30.72
7	Vietnam	0.32	7	Vietnam	0.37	15.09	7	Indonesia	0.29	−17.85
8	Indonesia	0.17	8	Indonesia	0.35	102.89	8	Vietnam	0.21	−41.53
9	Australia	0.15	9	Poland	0.31	15300.00	9	Poland	0.21	−32.14
10	Canada	0.12	10	Australia	0.21	40.27	10	Australia	0.11	−47.30
Total	16.40		Total	18.22		11.07	Total	13.17		−27.71

[z]$ million

[y](Current export value—past export value)/past export value × 100%

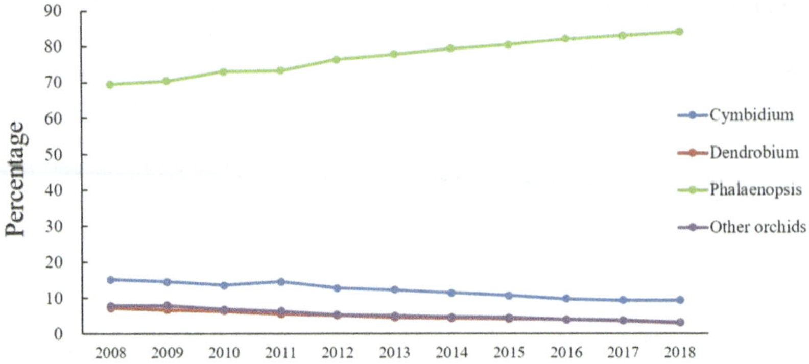

Fig. 1.10 Potted orchids in the Japanese wholesale market between 2008 and 2018. *Data* Central Wholesale Market, Japan (2018)

Dendrobium, extremely diverse in terms of color, size, and shape, is being considered as an emerging orchid genus in Taiwan. The total export value, including *D. nobile* hybrids and *D. phalaenopsis* hybrids, more than doubled, from $0.269 million in 2017 to $0.562 million in 2018, a 109% increase (Table 1.5). However, its export value declined by 63% to $0.211 million in 2019. *Dendrobium* orchids are mainly produced as potted plants. The biggest export market is Vietnam. While the value remained very low, exporting potted *Dendrobium* plants in growing medium to the United States increased in 2019 since the Animal and Plant Health Inspection Service (APHIS) of USDA amended its regulations in 2018 to allow for the importation of *Dendrobium* orchid plants in approved growing media from Taiwan into the United States.

The export value of *Cattleya* orchids was approximately $0.5 million, accounting for 0.3% of all orchids being exported (Table 1.6). Potted *Cattleya* plants were mainly exported to Vietnam. However, its share among the total value of the exported *Cattleya* gradually declined from 56% to 23% in recent years, probably due to decreased market demand.

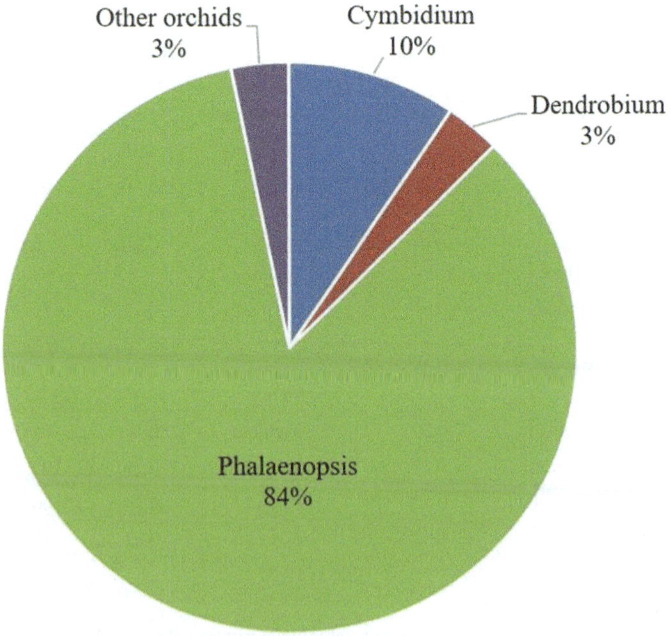

Fig. 1.11 The percentage of various potted orchids in the Japanese wholesale market in 2018. *Data* Central Wholesale Market, Japan (2018)

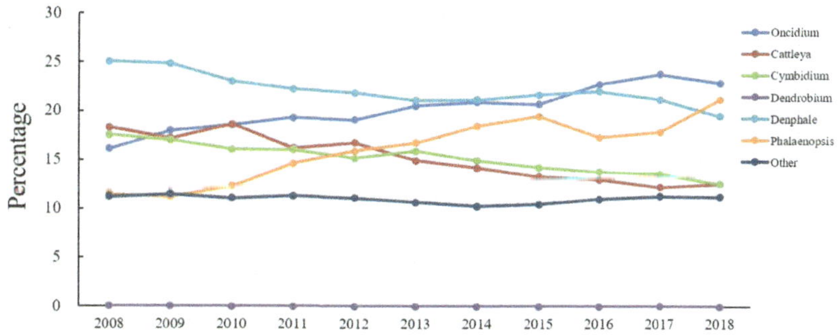

Fig. 1.12 Various orchid cut flowers in the Japanese wholesale market from 2008 to 2018. *Data* Central Wholesale Market, Japan (2018)

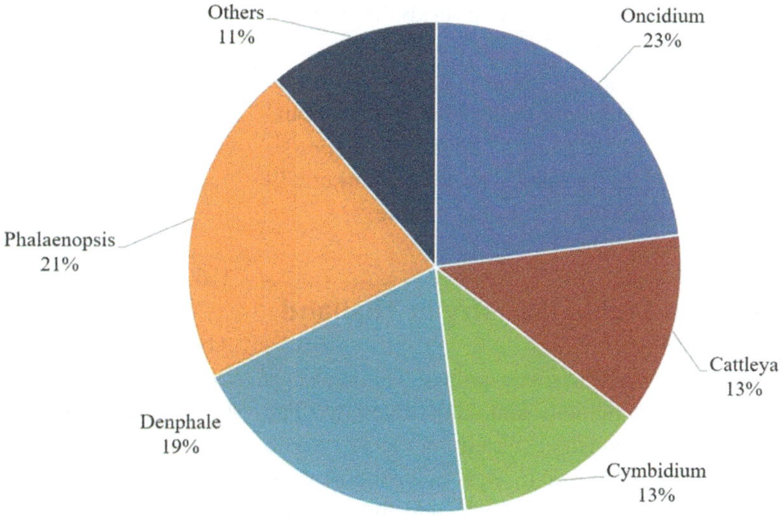

Fig. 1.13 The percentage of various orchid cut flowers in the Japanese wholesale market in 2018. *Data* Central Wholesale Market, Japan (2018)

The exported *Paphiopedilum* orchids, valued at $0.3 million in 2019 (Table 1.7), was exported mainly to Switzerland with a 144.5% growth, Canada with a 100% increase, United States with a 33.3% decline, Japan with a 54.6% reduction, and South Korea with a 87.5% drop over 2018. The drastically diminished export of *Paphiopedilum* orchids was probably due to their unpredictable flowering behavior and much shorter shelf life as compared to *Phalaenopsis*.

Taiwan also exported other unspecified orchid genera (Table 1.8), with 78–80% being exported to Japan. Interestingly, the Netherlands and South Korea imported more other orchid genera in 2019 as compared to 2018. On the contrary, the export of other orchids to China showed a 7.5% and 72.4% decline from 2017 to 2018 and from 2018 to 2019, respectively.

Exported flasks valued at $18.2 million, accounting for 9.2% of all orchids export in 2018, a 13.6% increase over 2017 (Table 1.9). However, in 2019, its export value declined by 27.7% over 2018. Most of the flasks were exported to the Netherlands.

1.5 The Japanese Wholesale Market

There are five major flower auction wholesale markets in Japan. Ota Wholesale Market had the largest total contract value in 2018.

In 2018, potted orchids collectively were the main floral commodity in the Japanese wholesale market, accounting for 61% among all flowers. The potted orchids in the Japanese wholesale market include *Cymbidium*, the nobile type *Dendrobium*, *Phalaenopsis*, and others (Fig. 1.10).

The demand for *Phalaenopsis* potted plants has increased in the Japanese floricultural market in recent years, while the demand for other orchids decreased. In 2018, potted *Phalaenopsis* had a market share of 84%, followed by *Cymbidium* at 10%, *Dendrobium* at 3%, and other orchids at 3% (Fig. 1.11).

Orchid cut flowers accounted for 39% of the Japanese orchid market with diverse product categories. Recently, the import of cut *Oncidium* and *Phalaenopsis* greatly increased, while cut *Dendrobium phalaenopsis* hybrids (Denphal), *Cymbidium*, and *Cattleya* decreased (Fig. 1.12). In 2018, the cut *Oncidium* had a 23% share, followed by *Phalaenopsis* at 21%, and Denphal at 20% (Fig. 1.13). It looks like the *Phalaenopsis* cut flowers in the Japanese wholesale market will exceed cut *Oncidium* in product value.

1.6 The Orchid Industry in Thailand

Orchids are found in every part of Thailand in a wide range of habitats and high diversity. The Thai orchid industry started in 1931, when *Dendrobium* Pompadour was first imported to Thailand. The plant adapted to and thrived well in Thailand's tropical climate. Later, it was grown commercially for cut flower. In 1957, the first 350 stems of *D.* Pompadour were exported. In 1984, *D.* Sonia, also known as BOM (Bangkok Orchid Mericlone), became a very popular cut flower cultivar. Many new cultivars derived from mutation through tissue culture such as *D.* Sonia 'Earsakul', *D.* Sonia '17 Daeng', *D.* Sonia 'Daeng Pimon', etc. Thailand is an important exporter of cut tropical orchids (*Dendrobium*, *Mokara,* and *Vanda*) and plants of *Dendrobium* and *Phalaenopsis*.

Thai orchid growers are well-known for their keenness on breeding, scouting, and producing their orchids. Therefore, there are always new hybrids or new cultivars either from selection or grower's own breeding program being introduced to the market. Because of the suitable climate in Thailand, good production can be carried out with simple shade houses at a reasonable cost.

Domestic consumption of cut orchids is largely used for paying homage on religious occasions, especially Buddhism at temples and at home. Orchid flowers are used for decoration at weddings, in hotel lobbies, and at big government functions as a sign of freshness and glamour. They are also popular for home decoration and funeral. Orchid plants are sold locally and find their way to residential houses or being attached to tree trunks. *Dendrobium* hybrids and many Thai native orchids, such as *Aerides* and *Rhynchostylis*, are very popular of being attached to tree trunks with abundant inflorescences and a long-lasting flowering period. Orchid plants are also popular as an arrangement in baskets or containers and presented as gifts for various occasions.

1.6.1 Orchid Production in Thailand

About 97% of the cut orchid production area is in the central part of Thailand mainly in Nakhon Pathom, Samut Sakhon, Ratchaburi, Bangkok, and Nonthaburi Provinces. In 2019, the cut orchid production acrage was estimated by Office of Agricultural Economics of Thailand at 3,443 hectares with a production of 48,794 tons, of which 25,707 tons (53%) were used domestically and 23,087 tons (47%) were exported. There were approximately 1,605 orchid farms, averaged at 2.15 hectare/farm. A survey conducted by the Department of Agricultural Extension in 2014 found the majority of orchid genera grown for cut flower were *D. phalaenopsis* hybrids (86.4%), *Mokara* (10.5%), *Vanda* (1.4%), and *Ascocenda* (1.4%). *D.* Sonia accounted for 71% of all *Dendrobium* being grown.

Dendrobium orchids produce flowers year-round with peak production from August to October and the lowest production from March till May which is one factor that causes the fluctuation in market price. Orchid plants for cut flower production have life cycles of 4–6 years.

Orchid plant production was estimated at 250 hectares with approximately 500 farms scattered around the country. A wide range of orchids

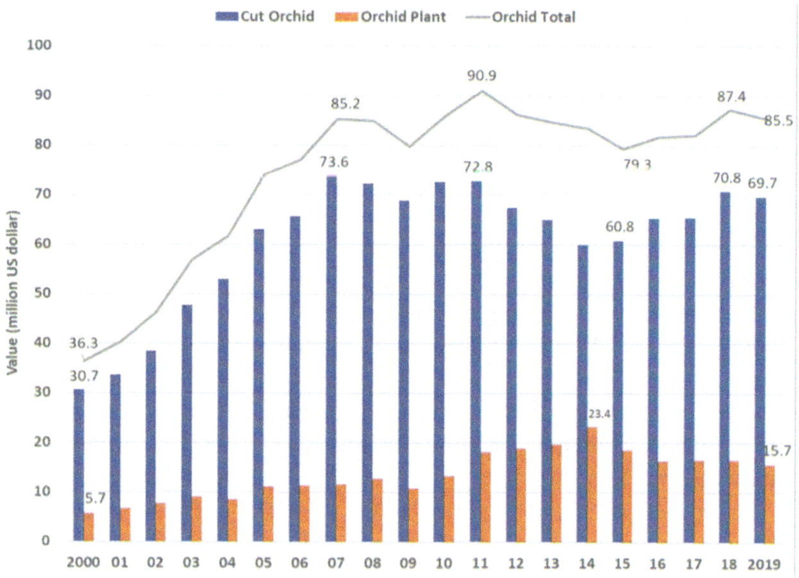

Fig. 1.14 Thai orchid export value from 2000 to 2019. *Data* Office of Agricultural Economics, Ministry of Agriculture and Cooperatives, Thailand

were produced for potted plants, the majority of which being *Dendrobium* and *Phalaenopsis*, followed by *Aranda*, *Cattleya*, *Oncidium*, *Rhynchostylis*, *Vanda*, and various native genera.

Orchid propagation is done mainly by tissue culture and division. There are many qualified tissue culture laboratories in Thailand that service Thai growers, as well as growers abroad.

1.6.2 Exports of Thai Orchids

In 2019, orchids exported from Thailand valued at $85.5 million, with $69.7 million or 81.6% from cut orchids and $15.7 million or 18.4% from orchid plants. The 20-year averaged annual growth rate was 3.1% for cut orchids and 6.0% for orchid plants (Fig. 1.14).

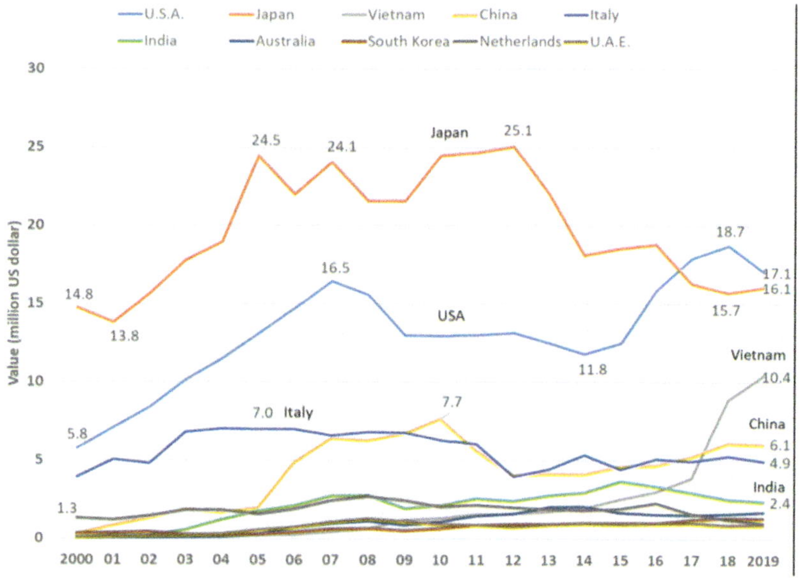

Fig. 1.15 Thai cut orchid export value in major importing countries from 2000 to 2019. *Data* Office of Agricultural Economics, Ministry of Agriculture and Cooperatives, Thailand

The largest export destination for Thai cut orchids in 2019 was the United States at $17.1 million and a 24.5% market share, with a 3.9% average annual growth for the past 10 years (Fig. 1.15). The Japanese market for Thai cut orchids was valued at $16.1 million and a 23.1% market share, at an averaged 5.6% annual decline for the past 10 years. Vietnam has had the fastest averaged annual growth rate of 25.2% for the past 10 years, with an import value of $10.4 million and a 14.9% market share. Other important customers were (in the order of import value) China, Italy, India, Australia, South Korea, the Netherlands, and the United Arab Emirates (Fig. 1.15).

Thai cut orchids were exported as (1) stems, (2) loose blooms, (3) leis, and (4) bouquets. Most cut orchids were exported as stems. In 2019, 505.9 million stems of cut orchids were exported. *Dendrobium* was the most popular genus exported with a 97.3% market share, followed by (in the order of value) *Mokara*, *Oncidium*, *Aranda*, *Aranthera*, *Vanda*, *Phalaenopsis*, *Rhynchostylis*, *Renanstylis*, and *Ascocenda*. Thailand exports cut orchid stems to China by container trucks which is more cost effective than by airplane; therefore, China imported the largest quantity of cut orchid stems (34.6% of all exports) (Table 1.10).

In 2019, 182.6 million pieces of orchid loose blooms were exported. Nearly all of the loose blooms were *Dendrobium* with only a few *Aranda* and *Mokara* blooms. They were imported by the United States (72.6%), India (15.5%), Japan (8.2%), Malaysia (7.4%), and South Korea (3.0%). In 2019, 4.7 million leis (garlands) were exported, nearly exclusively to the United States (98.6%). Almost all leis (99.95%) were made from *Dendrobium* flowers (Table 1.10).

Thai orchid plants were exported as (1) both young and mature plants, (2) flasks or plantlets in flasks which are grown in vitro from clonal propagation or seeds, (3), seedlings, and (4) bulbs. In 2019, 31.78 million orchid plants of more than 202 genera were exported from Thailand to 106 countries around the world. Among the genera exported, 68.8% was *Dendrobium*, 18.9% *Phalaenopsis*, 3.7% *Vanda*, and

Table 1.10 Quantity of Thai cut orchid products exported in 2019. *Data* Dept. of Agriculture, Ministry of Agriculture and Cooperatives, Thailand

Products name (Quantity/Growth 17–18)	Genera (share %)	Partner country (share %)
1. Stems (505.9 mil. stems/−6.8%)	*Dendrobium* (97.3%), *Mokara* (2.2%), *Oncidium* (0.18%), *Aranda* (0.18%), *Aranthera* (0.13%), *Vanda* (0.07%) *Phalaenopsis* (0.02%) *Rhynchostylis* (0.01%) *Renanstylis* (0.003%) *Ascocenda* (0.001%)	China (34.6%) Vietnam (18.3%) Japan (15.1%) India (8.1%) Italy (5.4%) USA (4.2%)
2. Loose Bloom (182.6 mil. pcs/−5.0%)	*Dendrobium* (99.9%) *Aranda* (0.01%) *Mokara* (0.003)	USA (72.6%) Japan (8.2%) Malaysia (7.4%) India (5.2%) South Korea (3.0%)
3. Lei (Garland) (4.7 mil. pcs/−1.6%)	*Dendrobium* (99.95%) *Mokara* (0.05%)	USA (98.6%) Japan (1.1%)
4. Bouquete (0.33 mil. pcs/−8.3%)	*Dendrobium* (100%)	Italy (69.4%) Australia (29.9%) Canada (0.37%) USA (0.35%)

3.6% *Rhynchostylis*. The major importers of Thai orchid plants were Vietnam (26.3%), Brazil (14.4%), Germany (11.4%), the Netherlands (10.8%), India (9.2%), the Philippines (6.0%), Indonesia (5.4%), and Sri Lanka (5.3%) (Table 1.11).

In 2019, 472,746 orchid flasks of more than 128 genera were exported from Thailand to 44 countries. The main genera exported were *Dendrobium* (42.1%), *Phalaenopsis* (17.8%), and *Oncidium* alliance (13.1%). The United States (27.9%), the Netherlands (16.7%), Vietnam (14.5%), and Malaysia (12.5%) were the major importers of Thai orchid flasks (Table 1.11).

The advantage of importing mature and ready to flower plants from Thailand is that plants are

Table 1.11 Quantity of various Thai orchids exported in 2018. *Data* Dept. of Agriculture, Ministry of Agriculture and Cooperatives, Thailand

Products name (Quantity/Growth 2018–198)	Genera (share %)	Partner country (share %)
1. Plant (31.78 mil. plts/ −17.4%)	***Dendrobium* (68.8%)** ***Phalaenopsis* (18.9%)** *Vanda* (3.7%) *Rhynchostylis* (3.6%) *Oncidium* (1.5%) *Mokara* (1.1%) *Cattleya* (1.0%) *Cymbidium* (0.3%) *Epidendrum* (0.2%) *Ascocenda* (0.1%) *Tolumnia* (0.1%) *Renanthera* (0.1%) *Spathoglottis* (0.1%)	**Vietnam (26.3%)** **Brazil (14.4%)** **Germany (11.4%)** **Netherlands (10.8%)** India (9.2%) Philippines (6.0%) Indonesia (5.4%) Sri Lanka (5.3%) Japan (2.0%) USA (1.9%) Dominican Republic (1.3%) Taiwan (1.3%)
2. Flask (0.47 mil. flasks/−38.2%)	***Dendrobium* (42.1%)** ***Phalaenopsis* (17.8%)** *Oncidium* (13.1%) *Rhynchostylis* (5.7%) *Cattleya* (5.6%) *Miltonia* (3.7%) *Cymbidium* (2.7%) *Mokara* (2.1%) *Zygopetalum* (1.4%) *Paphiopedilum* (1.0%) *Epidendrum* (0.7%) *Vanda* (0.6%) *Miltassia* (0.6%) *Burrageara* (0.4%) *Miltoniopsis* (0.3%) *Bulbophyllum* (0.3%) *Cambria* (0.2%) *Aranda* (0.1%) *Grammatophyllum* (0.1%) *Miltonidium* (0.1%)	**USA (27.9%)** **Netherlands (16.7%)** Vietnam (14.5%) Malaysia (12.5%) Brazil (9.3%) India (8.8%) Singapore (2.7%) Australia (2.5%) Japan (1.2%) Indonesia (0.8%) Reunion (0.8%) Taiwan (0.7%) Mexico (0.2%) South Africa (0.2%) Germany (0.2%) Russia (0.2%)
3. Seedlings (0.99 mil. seedlings)	*Brassia* (62.0%) *Dendrobium (35.3%)* *Phalaenopsis* (1.3%) *Cattleya* (0.7%) *Oncidium* (0.5%)	Netherlands (62.0%) Dominican Republic (22.2%) USA (7.0%) Sri Lanka (3.2%) Philippines (2.9%) Japan (2.0%)
4. Bulbs (368 pcs/−20.7%)	*Habenaria* (78.8%) *Pecteilis* (19.8%)	Canada (84.2%) Taiwan (14.4%)

allowed to grow rapidly in the tropical environment and then be shipped to finish them in greenhouses in temperate areas for blooming. This means less energy cost, good quality crops, and shortened production time.

1.7 The Orchid Industry in the United States

The orchid business has a long history in the United States. For more than a century, there were numerous orchid nurseries from coast to coast in every state, particularly in California. While some were of substantial sizes, many others were just barely larger than backyard operations. The vast majority of these orchid businesses were catered to the hobbyists with species and unique hybrids, fetching relatively high selling prices and earning enjoyable profits. Most of the orchids being sold were raised from seeds or from divisions. Another sizable business for the larger wholesale orchid operations was in the cut flower and corsage business, mainly *Cymbidium* and large-flowered *Cattleya* orchids, for holidays and graduations. Although a few are still in business, most of these commercial orchid nurseries with long and glorious histories have closed their doors for good over the past decades, particularly in the last 20 years as the hobbyist population dwindles quickly and the new generations buy orchids primarily at grocery stores.

Listed below are some of those well-known orchid nurseries. Among them, only three are still in the hands of the original owners. The others are in the memory of few orchid hobbyists and most of whom are in their senior years.

Armacost and Royston, Inc. Carpinteria, California
Carter and Holmes, Newberry, South Carolina
J. W. Bianchi Inc., Patchogue, New York
Dos Pueblos Orchid Company, Goleta, California
Gallup and Stripling, Carpinteria, California
Orchids by Hausermann, Inc., Villa Park, Illinois
Rod McLellan Co., South San Francisco, California
Santa Barbara Orchid Estate, Santa Barbara, California
Stewart Orchids, San Gabriel, Inc., California
The Orchid Zone, Inc., Castroville, California
Thomas Young Orchids, Bound Brook, New Jersey

A look at the January 1980 (41 years ago) issue of the American Orchid Society Bulletin found 53 pages out of the 104 pages in this issue being advertisements from orchid nurseries and their allied suppliers, all catering to the hobbyists. Today, most are out of business and few are still advertising for business. The American Orchid Society had over 45,000 members worldwide in the 1990s. Today, only thousands remain.

Prior to the early 1990s, there were no potted orchids on the supermarket shelves. Depending on the season of the year and holidays, cut flower sprays and orchid corsages were displayed for sale. Located over 10,500 km away to the west, Taiwan played a pivotal role in the establishment of the popular orchid industry in the United States in the modern history. In the summer of 1990, Taiwan Sugar Co., as part of its strategic move to diversify its business beyond sugar and hogs, began exporting *Phalaenopsis* orchids to the United States, a few thousand plants in that year. Due to the US agriculture quarantine regulations, *Phalaenopsis* orchids had to be removed of their growing medium (sphagnum moss) and imported to the United States as bare-root plants by air freight (Fig. 1.16). They were then potted into mostly 15 cm pots for recovery and flowering. That same year, a scientist at Texas A&M University initiated a research program that was dedicated largely to systematically study raising the imported *Phalaenopsis* as a new commercial crop en mass for the US consumers. The initial studies demonstrated the feasibility of importing the bare-root *Phalaenopsis* orchids in late summer or early fall and have them in flower for sale the following spring in 6–7 months of time.

During the few years ensued, techniques were quickly developed by Texas A&M University on potting medium selection, fertility, light requirement, conditions to induce and regulate flowering, regulation of flowering, postharvest handling, etc. to expand the production. Butterfly Orchids in Harlingen, Texas and Nurseryman's

Fig. 1.16 *Phalaenopsis* orchids being processed for export in a bare-root state

Exchange Inc. in Half Moon Bay and Rod McLellan Co. in South San Francisco, California were some of the pioneers to import bare-root *Phalaenopsis*. The strategy was to move away from the high expenses of growing *Phalaenopsis* from flask plantlets in the United States and lower the production costs by importing the cheaper mature plants and force them to bloom quickly for sale. All of the *Phalaenopsis* being imported then were seedling plants. Blooming orchids were only available during the spring months. With the development of research-based production technologies, blooming *Phalaenopsis* orchids quickly became available year-round. Grocery store chains began carrying potted blooming *Phalaenopsis* in the second half of the 1990s.

In 1997, due to the increased production and value of potted orchids, the United States Department of Agricultural (USDA) began tracking their volume and wholesale value as part of its annual floriculture survey. The data show a steady increase in popularity of orchids since 1996, the year the data first became available (Fig. 1.17). There was a staggering period between 2006 and 2008, but then the production kept climbing. The USDA data show, in 2019, growers shipped out close to 39 million pots of orchids at a wholesale value of $308 million, including the state of Hawaii. The majority of the potted orchid being produced today for the popular market are clonally propagated *Phalaenopsis*.

In 2018, Hawaii produced 1.5 million pots of orchids at a wholesale value of $12 million. Therefore, when those from Hawaii are added to the total production and sales of orchids on the mainland, United States produced over 36.2 million pots of orchids that valued at $305 million in 2018. Hawaii growers produced $3.4

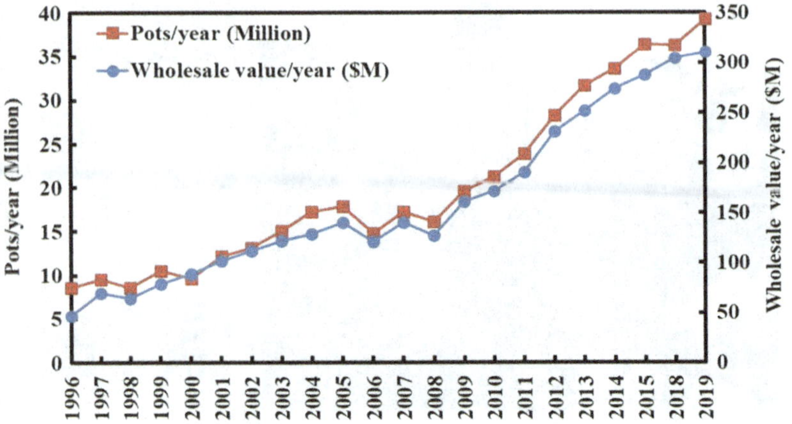

Fig. 1.17 The production volume and wholesale value of potted orchids in the United States, including the state of Hawaii, 1996–2019. *Data* The United States Department of Agriculture, NASS (1997)

million worth of cut orchids in 2018, the last data available.

In 2005, the USDA approved the importation of *Phalaenopsis* orchids from Taiwan in their original growing pots with the existing growing substrate (nearly all in sphagnum moss). Technologies were developed to ship *Phalaenopsis* orchids to the United States in ocean reefers on container ships (Fig. 1.18). Orchid plants are mostly packed in cartons, but some may be packed on multilayer carts. Depending on the port of docking, it takes two weeks to Long Beach, California and four to five weeks to New York. Upon arrival, pots are immediately placed on greenhouse benches and air temperature is dropped to induce spiking (the emergence of the flowering stems) without a “recovery period”. Depending on the variety, *Phalaenopsis* orchids are available for sale in 12–21 weeks following their arrival. Usually, a greenhouse produces two crops per year.

Due to the substantial lower freight costs of orchid plants as a result of shipping by the sea, the quantity of *Phalaenopsis* orchids being exported from Taiwan to the United States increased quickly between 2006 and 2014 (Fig. 1.19). The growth rate slowed down since 2015. The wholesale value of potted orchids in the United States first exceeded that of poinsettia in 2009 to become the most valued potted floral crop in the United States, with a respective value of $160 and $145 million. Currently, Matsui Nursery in California is the only large orchid operation that produces a significant amount of orchids other than *Phalaenopsis*, such as *Dendrobium*, *Miltoneopsis*, as well as *Oncidium* and its intergeneric hybrids.

In 2016, potted *Oncidium* orchids from Taiwan and *Phalaenopsis* from the Mainland China were approved for importation to the United States. *Phalaenopsis* and *Cymbidium* in their original growing container were approved for importation from South Korea in late 2017. In 2018, USDA approved the shipment of *Dendrobiums* from Taiwan in their growing pots. As of early 2021, there is no other country or region that has been approved to export orchids in their original growing medium to the United States. However, APHIS of USDA has completed risk assessments in 2020 to potentially allow the importation of *Phalaenopsis* from Germany (no limit, young and mature plants) and the Netherlands (5 million young plants per year) in approved media.

Another production system took hold in the United States since the early 2000s. Several nurseries added *Phalaenopsis* to their existing product line by adopting the Dutch system of growing *Phalaenopsis* in a pine bark mix and with high degrees of automation. These nurseries included Green Circle Growers, Mid-American Growers, Silver Vase, Westerlay Orchids, etc. Floricultura, the world’s largest *Phalaenopsis* propagator and young plant supplier, began its operation in the Central Coast of California at the end of 2010 (Fig. 1.20). After the completion of

Fig. 1.18 *Phalaenopsis* are shipped in temperature-controlled (~18 °C) ocean containers from Taiwan to the United States. They are subject to additional quarantine inspection by the USDA at their ports of entry before being released to the local nurseries

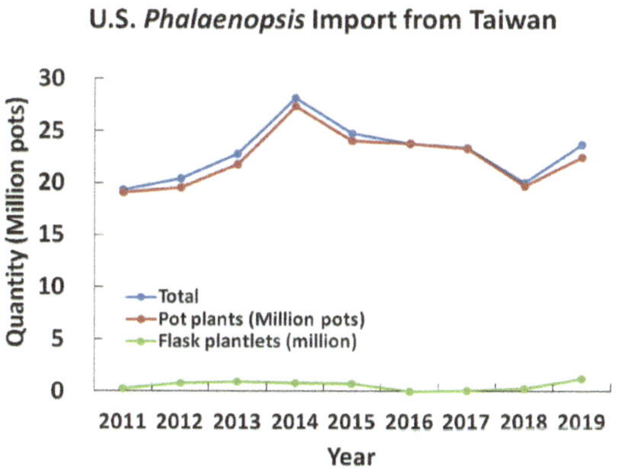

Fig. 1.19 The quantity of the *Phalaenopsis* exported from Taiwan to the United States. *Data* Customs Administration, Ministry of Finance, Taiwan

additional greenhouses in 2016, production capacity reached 7 million pots annually. Altogether, it is estimated that these orchid nurseries produce about 15–16 million pots a year.

On the US consumer market, among the 39 million pots of orchids being produced in 2019, it is estimated that 55% of these orchids are planted in sphagnum moss, while 45% is produced in bark mixes. Due to the high water-holding capacity of the sphagnum moss, orchids grown in moss are irrigated much less frequently than those planted in a bark mix.

Orchid production on the continental United States concentrates mainly on the two coasts (Fig. 1.21). California and Florida each

Fig. 1.20 *Phalaenopsis* at Floricultura Pacific (left) and a variety of orchids being prepared for shipping at Matsui Nursery (right). Both are outside of Salinas, California on the Central Coast with a unique cool weather year-round

Fig. 1.21 The major orchid-producing areas in the United States. The size of the orchid flowers on the map represents the estimated value of the production for that given area. Numerical data for California and Florida: US Department of Agriculture, NASS, Floriculture Crops 2019 Summary, 2020. Others, estimates

accounted for 45% and 30%, respectively, of the total national value in 2019 (including Hawaii estimated at $12 million for 2019). Since the legalization of *Cannabis* in California (2018) and Florida (2016), some major orchid operations have been transformed to producing this lucrative crop. *Cannabis* does not appear to have negatively affected orchid production in the United States (Fig. 1.17).

1.7.1 Orchid Production in Hawaii

The floriculture and nursery industry is an important sector of diversified agriculture in Hawaii. In 2018, the wholesale value of floriculture and nursery products was estimated at $75.68 M (NASS 2019). Hawaii produces a variety of orchids used as individual blossoms, cut flowers, and flowering potted plants. Due to the varying microclimate in various areas of the state, dendrobium and the vandaceous orchids (*Aranda*, *Aranthera*, *Mokara*, and *Vanda*) are grown in locations with higher light intensities, warmer weather, and relatively dry conditions. *Cattleya*, members of the *Oncidiinae* subtribe (*Oncidium*, *Brassia*, *Miltonia*, *Miltoniopsis*, and intergeneric hybrids such as *Colmanara*, *Degarmoara*, and *Odontocidium*), *Paphiopedilum*, and *Phalaenopsis* are grown in areas with moderate sun and slightly cooler conditions. *Cymbidium* is grown in higher altitudes (Gilliam and Hiranaga 2017).

The wholesale value of orchids produced in the State of Hawaii from 2000 to 2018 increased from 2000 to 2003 (Fig. 1.22). Wholesale value peaked in 2003, with sales of $23.4 million, and was sustained above $20 million until 2007. In 2008, the wholesale value of Hawaii orchids dropped by 16% and another 13% in 2009, coinciding with the worldwide economic downturn that started in December 2007 that lasted until June 2009. Growers attributed the decline in 2008–2009 to the sluggish economy, crop losses to volcanic emissions, adverse weather conditions, and the high cost of farming, particularly chemicals and fertilizer prices. In addition, some plant exporters reported that restrictions on Hawaii exports limited their sales (HI NASS 2009). After steady declines from 2010 to 2011, wholesale value began to modestly increase. However, in August 2014, Hurricane Iselle hit Hawaii Island, damaging production areas. Consequently, 2015 saw a decline in wholesale value. In May 2018, the Kilauea East Rift zone eruptions began on the island of Hawaii. Lava flowing over the orchid production areas in Kapoho Bay resulted in 100% destruction of the Kapoho nurseries (Loke 2018; Inouye 2020). The repercussions from the lava flow are expected to be reflected in the 2019 sales value and production statistics.

In Hawaii, cut flower production is focused on *Dendrobium*, the *Oncidiinae*, *Cymbidium* and other orchids (Fig. 1.23). Cut orchid sales decreased from $3.3 million in 2000 to $1.7 million in 2018. Percentage of sales from cut flower orchids has been less than that of potted plant sales. From 2000 to 2008, cut flower sales comprised 18–22% of Hawaii orchid sales. Following the economic downturn in 2008–2009, percentage of cut flower sales dropped to 16% in 2009. Modest gains were realized from 2012 to 2013, increasing market share to 22% in 2012 and 21% in 2013. In 2018, cut flower orchid sales comprised 13% of the total orchid sales in Hawaii.

Of the orchid groups, the volume and value of cut dendrobium have been considerably higher compared to the other orchid groups, as high as 78% of cut orchid sales in 2001, and gradually diminishing to 60% in 2007, followed by another drop to 59% in 2010 and 2011, but rebounding to 74% in 2012 and 75% in 2013. Since then, cut dendrobium demand has dropped to a steady 60% range. According to the Orchid Growers of Hawaii, the association of commercial orchid

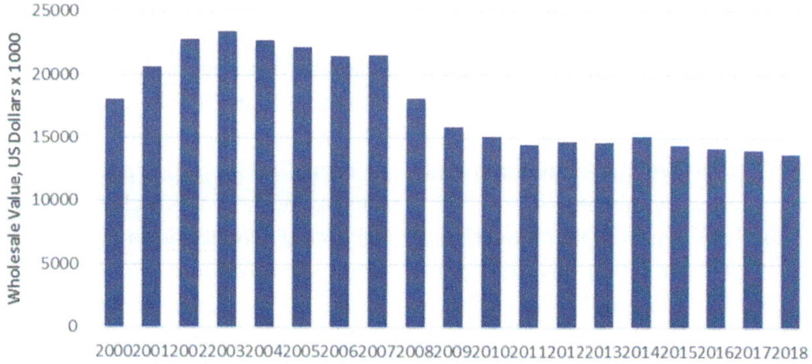

Fig. 1.22 Wholesale value of orchids produced in the State of Hawaii, 2000–2018. *Data* USDA-NASS Hawaii Floriculture and Nursery Products Annual Summary

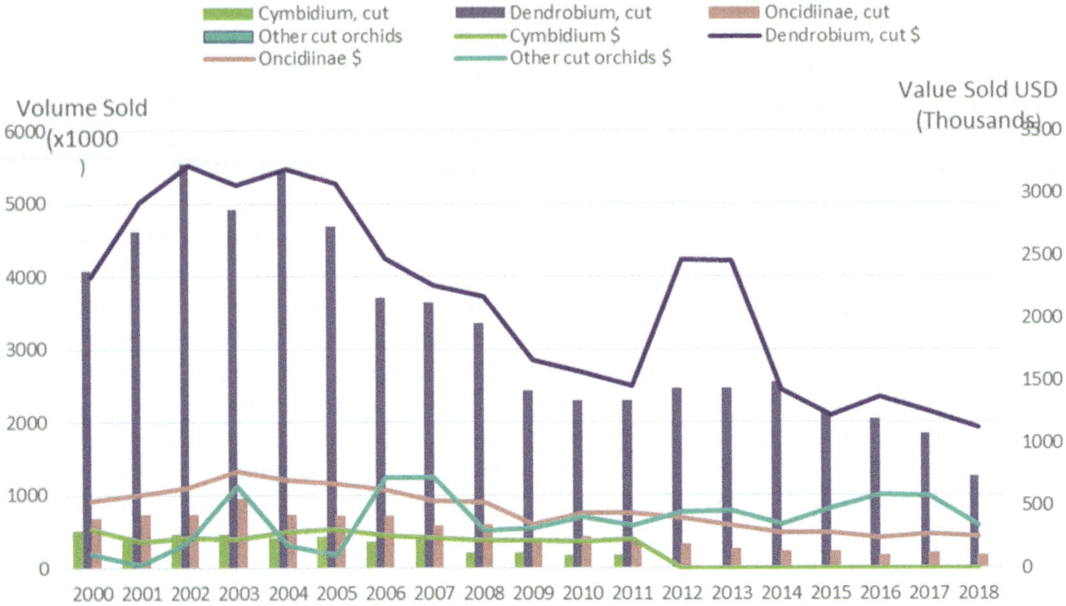

Fig. 1.23 Cut flower orchid production in Hawaii, 2000–2018. *Data* Summarized from the Hawaii Flower and Nursery Reports 2000–2018, National Agricultural Statistics Service

growers in Hawaii, current demand is good in the marketplace especially in the winter months and holidays and a surplus in summer months (J. Tanouye, personal communication). The small flowered intersectional hybrids between *Phalaeananthe* and *Spatulata* sections, exemplified by hybrids of *Dendrobium* Jaquelyn Thomas are unique to Hawaii. These hybrids have been developed at the University of Hawaii and are propagated by seed (Kamemoto et al. 1999). Their exceptional stem length with numerous blossoms makes them a premium choice of designers for wedding and party work (Gilliam and Hiranaga 2017). White, purple, pink, green, blush are popular colors for cut dendrobium. *D.* Jaquelyn Thomas 'Uniwai Mist' (also known as UH800) is still the most produced variety due to its high yield and stem length (J. Tanouye, personal communication) that are preferred by customers.

Cut *Cymbidium* is grown in higher elevations. However, since 2012, production has been considerably less, such that the production statistics have not been included in the annual reports since 2012. The main *Oncidiinae* cut flowers include *Oncidesa* (*Oncidium*) Gower Ramsey, *Colmanara* Wildcat, *Degarmoara* Toy Soldier 'Volcano Queen', and *Odonticidium* Catatante (Gilliam and Hiranaga 2017). Other cut orchids include hybrids of *Vanda*, *Phalaenopsis*, and *Paphiopedilum*. The vandaceous cut orchids cultivated include *Aranda* Chao Praya 'Dot Com', *Aranda* Noorah Alsagoff, *Aranthera* Azimah, *Aranthera* Kapoho Fire, *Mokara* Chao Praya, and *Mokara* Dinah Shore (Gilliam and Hiranaga 2017).

The market share of cut *Oncidiinae* has been steady from 2000 to 2018, ranging from 10% in 2013 to 18% in 2011. In 2018, cut *Oncidiinae* comprised 15% of cut orchid sales. The market share of "other cut orchids" increased from 4% in 2000 to a high 28% in 2017 and then dropped to 20% among all cut orchids in 2018.

From 2000 to 2018, potted orchids constituted roughly 80% of the orchid sales in Hawaii. Potted orchid production in Hawaii includes hybrids from *Dendrobium*, *Phalaenopsis*, and other genera (Fig. 1.24). Included in other potted orchids are *Cattleya* and allied genera, *Oncidiinae*, *Paphiopedilum*, and *Vanda* and its allied genera (Gilliam and Hiranaga 2017). Of the potted orchids grown in Hawaii, *Dendrobium* used to be the predominant genera. However, production of other potted orchids (primarily *Oncidiinae*) has surged in volume and value over the past decade. In 2000, *Dendrobium* constituted 44% of potted orchid sales with wholesale value of $6.5 million, while other potted orchids contributed $7.2 million, about 47% of potted orchid sales. By 2018, potted *Dendrobium* sales totaled $2.5 million or 21% of potted orchid sales, while sales of other orchids were valued at $7.3 million or 61% of potted orchid sales in Hawaii.

The listing of the top 100 hybridizers producing the most crosses that originated from 2000 to 2018 (Note: acknowledged as the originator or hybridizer when the new hybrid was registered with the Royal Horticultural Society in the United Kingdom) was obtained for each of the major orchid genera (OrchidWiz Database 6.2, 2020). Fourteen hybridizers of *Oncidiinae* from Hawaii were credited for originating 733 hybrids from 2000 to 2018. Of the 14, three were ranked in the top 10 originators of *Oncidiinae* hybrids from 2000 to 2018. Hawaii hybridizers produced a total of 766 *Oncidiinae* hybrids during 2000–2018.

The number of orchid farms has been decreasing from 2000 to 2018 (Fig. 1.25). Many of the orchid farms have closed or have since downsized. According to the Orchid Growers Organization of Hawaii (J. Tanouye, personal communication), not many new, young growers are joining the orchid production industry. To attract new growers, the industry organization initiated partnerships with educational institutions, including community colleges that grant degrees within two years or colleges which grant bachelor's degrees, to encourage students to explore careers in plant production. Increasing production costs associated with labor (increases in minimum wage and small pool of workers willing to work on farms), raw material (cinder, rocks, water, fertilizer), new pests, and diseases are challenges facing growers in Hawaii.

1.7.2 Effects of the 2018 Volcanic Eruption

In May 2018, the Kilauea East Rift zone eruptions began on the island of Hawaii where the majority orchid producers are located. Volcanic emissions attributed to hydrogen sulfide caused

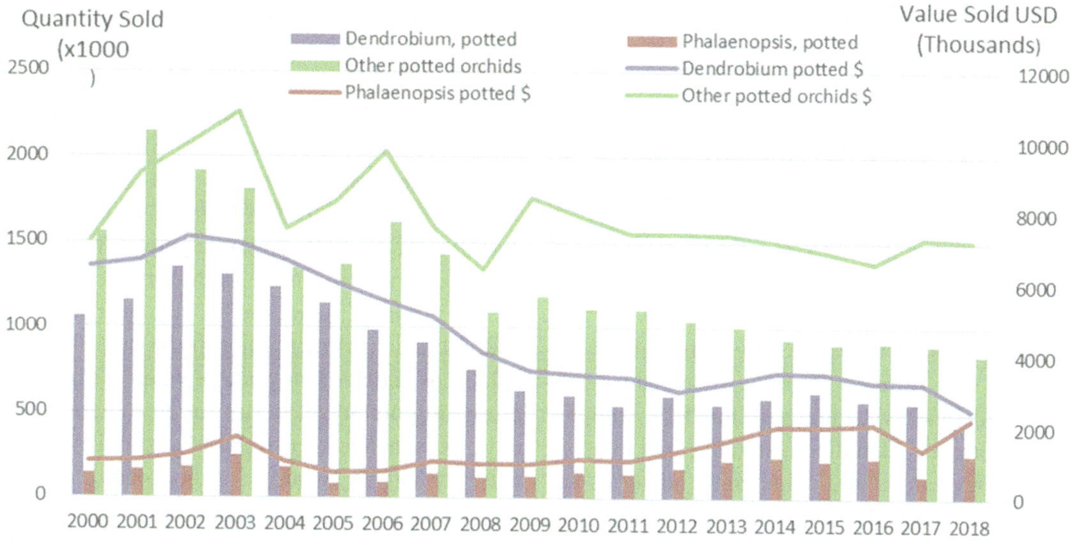

Fig. 1.24 Potted orchid production in Hawaii, 2000–2018. *Data* Summarized from the Hawaii Flower and Nursery Reports 2000–2018, National Agricultural Statistics Service

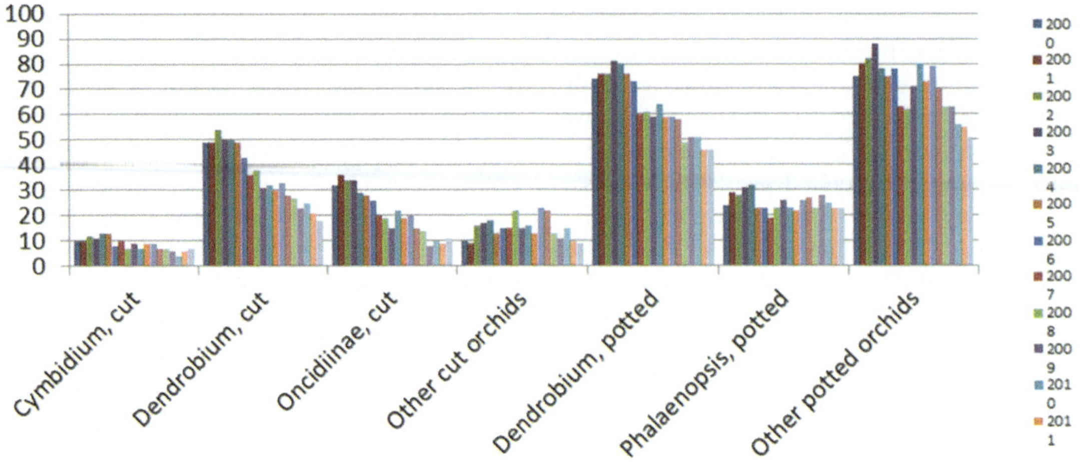

Fig. 1.25 Number of farms in Hawaii with sales of over $100,000 annually, by orchid genera, 2000–2018. *Data* USDA-NASS Hawaii Floriculture and Nursery Products Annual Summary

leaf drop in dendrobium plants in production. In June 2018, lava flowed over the orchid production areas in Kapoho Bay (Inouye 2020) resulting in total loss for some growers and loss of road access for virtually all growers. Shortly after, a farm disaster survey was conducted to assess the extent of damage brought about by the volcanic emissions and lava flow (Loke 2018). The survey revealed that total farm losses due to the eruption were $27.9 million. Crop losses amounted to $17 million or 61%. Total loss to floriculture and nursery sector was estimated at $13.3 million, highest loss of crop value in the area. Crop loss was estimated at $7.5 million, while land loss was estimated at $1 million. Damage to building structures was estimated at $2.3 million, while inventory losses were estimated at $2.5 million. Most inundated farmers were willing to start over or acquire new farm but few were willing to expand their size.

A partnership between Hawaii Island orchid growers who suffered property losses and a private landowner was initiated in 2019 (Inouye 2020). Growers identified a suitable area to lease so that they can rebuild commercial nurseries for growing orchids. Land preparation and greenhouse construction have started. It will take some years before cut flowers can reach production stage. The future of the Hawaii orchid industry is still unfolding.

1.8 The European Orchid Market and Upcoming Consumer Trends

The orchid has taken a significant position in the pot plant industry due to its distinct qualities: attractiveness and long shelf life. Due to the current market, developing orchids and selling them is not what it used to be. The market is not the same and consumers aren't the same anymore. With the development of online E-commerce platforms, the sales, marketing, and branding approaches are different from before, and apparently will have more market share during the Covid-19 pandemic period. We aim to share opportunities in the orchid market.

For instance, social media, digital marketing, and multichannel marketing in the past were not valued as much as nowadays when it comes to promoting orchid sales. But for customers nowadays, personalized services and customized products are highly valued in the orchid market. Especially for those middle to high-end customers, the branding strategy needs to change and go with the time.

Orchids account for a large share of the global floriculture trade as potted orchids are estimated to comprise more than 10% of the international pot plant trade. And among the markets worldwide, the Americas, the EMEA, and Asia are the most important and active ones.

1.8.1 The Customer in 2030

For the next decade, both customer mindset and behaviors will influence their decision of purchasing orchid, so we need to further examine the correlation between personalized lifestyle and their willingness to purchase orchids. For example, the orchid may symbolize a personal taste or living style; therefore, if certain image can be linked to the orchids, more customers will be attracted into the orchid market. Besides, how to maintain those existing customers and how to keep them satisfied and stay loyal with orchids are also challenging and further considerations and research are required. The most important part about this trend is to know your customer and where do they put the orchids after the purchase. These two variables are important for the orchid industry to continue their relevance in the European orchid market. Previous research also shows us relevant data regarding this topic. When gifted or purchased, 29% of the orchids usually end up in the family room and 28% is displayed in the kitchen. This kind of data doesn't seem relevant, but it is extremely important to understanding the whole lifecycle of the orchid in the value chain and also in the customer homes to keep the industry alive.

1.8.2 Purchasing Motivators

Among purchasers, the most important factor in the decision-making process is "reasonable price". Once the price is acceptable, customers consider the following purchase factors jointly: offering caring instructions, longer blossom period, and minimal chemical residues. Also, one of the most important characteristics of the orchid is that it doesn't need much care or water. These characteristics can be very important variables in the branding and marketing of the orchid.

1.8.3 Orchid Market Statistics and the *Phalaenopsis*

According to Royal FloraHolland, *Phalaenopsis* has been the most important orchid crop in the world with a market share of 79% among all orchids being sold. Statistics from Royal FloraHolland showed that, in 2018, 130 million pots of *Phalaenopsis*, valued at €410 million, were sold in the Netherlands. Of the 130 million *Phalaenopsis*, 19% were sold through the Clock Auction and 81% were direct sales. A consistent 70% of the *Phalaenopsis* plants being sold had two or more inflorescences. The 12 million pots of other orchid genera, 40% of which being the nobile type *Dendrobium*, were €51 million in value. There were some 40 million pots of orchids in other European countries (Poland/Italy/Germany/Denmark/UK local producers), likely generating another €160 million or more at the wholesale level. Altogether, in addition to the cut orchids, the wholesale value for all of the 182 million potted orchids in Europe was more than €620 million in 2018.

The European market is considered very competitive with many producers, but there are opportunities to grow in the mini flower type, like the small pots and supplying the semi-finished products. In addition to smaller and hobby breeders in Taiwan, about the same size of the Netherlands, the diverse products provided by the Taiwanese vendors have established prestigious record internationally. They are also skillful and experienced in all stages of orchid production.

Phalaenopsis is Taiwan's predominant flower export, at 96% of the total flora export value in 2019 (Fig. 1.6). However, Taiwanese orchid products are also threatened by competitors, mainly the Netherlands. For example: the major *Phalaenopsis* breeders are based in Europe. The main growers are located in the Netherlands with other smaller growers in Belgium, Germany, and Poland. Growers in Europe usually start from young plants while most growers in North America often start from semi-finished plants.

Among the pot orchids, *Phalaenopsis* remains the top one in the FloraHolland auction market. The wholesale value for *Phalaenopsis* pot plants was €450 million in 2019 (Royal Floralland 2019), with a slight increase over 2018.

1.9 Conclusion and Perspectives

Orchids have become quite popular in many countries, with the global trade increasing steadily annually, despite the COVID-19 pandemic. More new orchid varieties are expected to be developed to fulfill the consumers' demand in the dynamically changing marketplace. Orchids with desirable traits, most importantly long shelf life, novel colors, and fragrance, should be available in the near future from Taiwan and the Netherlands, countries active in breeding. In addition, tissue culture laboratories are advancing their technology for shortening the production period, convenient shipment of flasks, and high production rate. The areas with suitable climate environment will become the base for supplying young orchid plants. Future global orchid market seems to have a trend of strong collaboration on breeding, tissue culture, young or finished plant production, and marketing. This would be the best supply chain for future orchid businesses worldwide.

References

1. Central Wholesale Market, Japan (2018) Agricultural statistic information. 2018. http://www.shijou-tokei.metro.tokyo.jp/asp/rmenu.aspx?gyoshucd=3. Retrieved 1. September, 2019
2. Council of Agriculture, Taiwan (2010) Agricultural statistic information. 2018. https://agrstat.coa.gov.tw/sdweb/public/trade/tradereport.aspx. Accessed 1 September, 2019
3. Customs Administration, Ministry of Finance, Taiwan (2019) Trade statistics search. 2019. https://portal.sw.nat.gov.tw/APGA/GA03E. Retrieved 1 October, 2019
4. Royal Flora Holland (2019) Annual Report, Retrieved: https://www.royalfloraholland.com/en/about-floraholland/who-we-are-what-we-do/facts-and-figures/annual-reports/annual-report-2019
5. Gilliam H, Hiranaga L (2017) Neotropica: Hawaii Tropical Flower + Plant Guide (2nd ed). Design358 Publishing, Vancouver, Canada
6. Inouye G (2020) Retrieved from: https://www.hawaiifloriculture.org/videos/updates-on-disaster-emergency-relief-projects-programs
7. Kamemoto H, Amore TD, Kuehnle AR (1999) Breeding Dendrobium Orchids in Hawaii. University of Hawaii Press
8. Loke M (2018) Farm disaster survey results: Kilauea East Rift Zone eruptions, 2018. Retrieved from: https://gms.ctahr.hawaii.edu/gs/handler/getmedia.ashx?moid=31033&dt=3&g=12
9. National Agricultural Statistics Service (1997) Hawaii Nursery & Nursery Products Annual Summary 1996. Retrieved from https://www.nass.usda.gov/Statistics_by_State/Hawaii/Publications/Flowers_and_Nursery_Products/Floriculture/xflo96.pdf
10. National Agricultural Statistics Service (2002) Hawaii Nursery & Nursery Products Annual Summary 2001. Retrieved from: https://www.nass.usda.gov/Statistics_by_State/Hawaii/Publications/Flowers_and_Nursery_Products/Floriculture/xflo01.pdf
11. National Agricultural Statistics Service (2006) Hawaii Nursery & Nursery Products Annual Summary 2005. Retrieved from: https://www.nass.usda.gov/Statistics_by_State/Hawaii/Publications/Flowers_and_Nursery_Products/Floriculture/xflo05.pdf
12. National Agricultural Statistics Service (2011) Hawaii Nursery & Nursery Products Annual Summary 2009. Retrieved from: https://www.nass.usda.gov/Statistics_by_State/Hawaii/Publications/Flowers_and_Nursery_Products/Floriculture/xflo09.pdf
15. National Agricultural Statistics Service (2015) Hawaii Nursery & Nursery Products Annual Summary. Retrieved from: https://www.nass.usda.gov/Statistics_by_State/Hawaii/Publications/Flowers_and_Nursery_Products/flower.pdf
16. National Agricultural Statistics Service (2019) Hawaii Horticulture and Nursery Products Annual Summary 2018. Retrieved from: https://www.nass.usda.gov/Statistics_by_State/Hawaii/Publications/Flowers_and_Nursery_Products/Floriculture/201909HawaiiWholeFlower.pdf
17. National Agricultural Statistics Service (2019) Floriculture crops. Retrieved from: https://usda.library.cornell.edu/concern/publications/0p0966899?locale=en

2 The Breeding of *Phalaenopsis* Hybrids

Shih-Chang Yuan, Pablo Bolaños-Villegas, Chin-Yi Tsao, and Fure-Chyi Chen

Abstract

The lush and exuberant subtropical forests of South East Asia are home to about 60 native species of *Phalaenopsis* orchids. Results from research into evolutionary and reproductive relationships within this genus are reported with the hope that their understanding may assist the selection of species as parents during breeding programs. The species in sub-genus *Phalaenopsis* have been used frequently for commercial breeding and trading. Polyploid cultivars with superior horticultural traits are the mainstream, and are the product of either the natural production of unreduced gametes or the result of artificial production during the process of cross hybridization. Meiocyte analysis may provide valuable information regarding the formation of unreduced gametes and subsequent use in breeding for polyploidization. Artificial induction of polyploidy may be brought about by chemical treatments or as an unintended result of mutations during the tissue culture process. Proper storage of orchid pollinia in sub-zero freezing temperatures may extend their viability until the moment they are required for cross hybridization. This chapter elaborates into these methods and strategies with the hope that they might contribute to new orchid breeding programs.

2.1 Introduction: *Phalaenopsis* Species and Their Importance in Breeding

There are more than 60 *Phalaenopsis* species, a monopodial epiphytic orchid, distributed in South East Asia (Lin 1977; Christenson 2001) with sub-tropical climates. The flowering season of *Phalaenopsis* species can be divided into two groups, spring and summer flowering (Lin 1977; Christenson 2001). The spring flowering species are usually cool temperature sensitive for spiking, while summer flowering species are temperature non-sensitive. In the Wikipedia about 70 species are described in this genus (https://en.wikipedia.org/wiki/Phalaenopsis). Several new species were described or newly discovered. Some species were reclassified from other

S.-C. Yuan · F.-C. Chen (✉)
Department of Plant Industry, National Pingtung University of Science and Technology, Pingtung, Taiwan
e-mail: fchen@mail.npust.edu.tw

P. Bolaños-Villegas
Fabio Baudrit Agricultural Research Station, University of Costa Rica, La Garita, Alajuela 20101, Costa Rica

Jardín Botánico Lankester, Universidad de Costa Rica, Cartago P.O. Box 302-7050 Costa Rica

C.-Y. Tsao
Section of Biotechnology, Taiwan Agricultural Research Institute, Wu-Feng, Taichung, Taiwan

F.-C. Chen and S.-W. Chin (eds.), *The Orchid Genome*, Compendium of Plant Genomes,
https://doi.org/10.1007/978-3-030-66826-6_2

alliance genera, such as *Phalaenopsis difformis* (synonym *Ornithochilus difformis*), *P. japonica* (synonym *Sedirea japonica*) and *P. marriottiana* (synonym *Hygrochilus marriottiana*). Around the world, horticultural markets often showcase potted *Phalaenopsis* orchids with large flowers, in addition to their more traditional use as cut flowers. *Phalaenopsis* hybrids can be obtained from intra-specific and inter-specific hybridization. Practically few commercial hybrids are species in its nature, while most hybrids are contributed by many species through generations of introgression by cross hybridization. An understanding of the species relationship in *Phalaenopsis* genus may help in understanding genetic closeness among different phylogenic groups so as to assist in the selection of desired species in a breeding program. The analysis of plastid and nuclear ribosomal DNA has allowed to conduct molecular phylogenetic comparisons in *Phalaenopsis* and related genera (Yukawa et al. 2005; Tsai et al. 2005, 2010). Results from the analysis of internally transcribed sequences (ITS) suggested that the *Phalaenopsis* section of within sub-genus *Phalaenopsis* was monophyletic (Tsai et al. 2005). Our own study based on chloroplast *matK* sequences, which are quite conserved across plant species, also revealed a similar pattern of monophyletic clustering of *Phalaenopsis* species. For that study we used the neighbor-joining method with two different bootstrap values (Fig. 2.1a, b) (Tsao 2003). The species in the *Phalaenopsis* section are mainly winter or spring flowering species which have contributed greatly to most of the modern commercial hybrids traded around the world, although species from other sections are also used for the breeding of spotted and waxy flowers.

Different markets (e.g. countries) may require unique traits in flowering potted plants of *Phalaenopsis*, such as spike number, plant height, floret number and size, inflorescence angle and branching. Specific traits such as flower size, colors like white, pink, indigo and violet, strong fragrance, lip morphology and heavy pigmentation spots (the harlequin pattern) could also be targets for breeding. Therefore, a successful breeding program needs to consider these market demands so as to select superior parents with certain required horticultural traits for cross hybridization. To obtain better or more advanced hybrids, the establishment of records that collect experience and database information (or pedigree records) must be a priority when one initiates a breeding program. A modern commercial hybrid may contain up to 15 or so species in its genetic background. Some commercial cultivars may have many desirable horticultural traits and can serve as superior parent for cross hybridization. One typical miniature cultivar, *Phalaenopsis* Sogo Vivien 'Sogo F858' (Fig. 2.2a), is immensely popular in the global market. Its ancestor can be traced back to about 11–12 generations of hybridization, and from the pedigree or genealogy of the hybrid it was revealed that a total of 11 species contributed certain traits to this cultivar through repeated hybridization and introgression (Fig. 2.2b). Another famous miniature cultivar, *P.* Liu's Twilight Rainbow 'Sogo F2006' (Fig. 2.3) was the result of a contribution by three species, namely *P. equestris*, *P. aphrodite* and *P. stuartiana*, through three generations of cross hybridization integration.

2.2 Breeding Strategy

The *Phalaenopsis* hybrid cultivars are rather diverse in their genetic constitution, and could be contributed by several species, or be an intra-specific hybrid in nature. When the flower is fully open, one can pollinate with pollinia of its own (self-pollination) or with pollen from other hybrids. Introgression means crossing progeny with one of its parents to obtain certain desirable horticultural traits. To satisfy current market trends for commercial cultivars, some species/cultivars be selected because they possess certain desirable traits that could be incorporated into current cultivars, such as multiple spikes, miniature flowers, color, fragrance and flower shape. Thus, a hybrid can be inter-crossed to another hybrid, or to a species. In brief, the breeding strategies that may be employed are

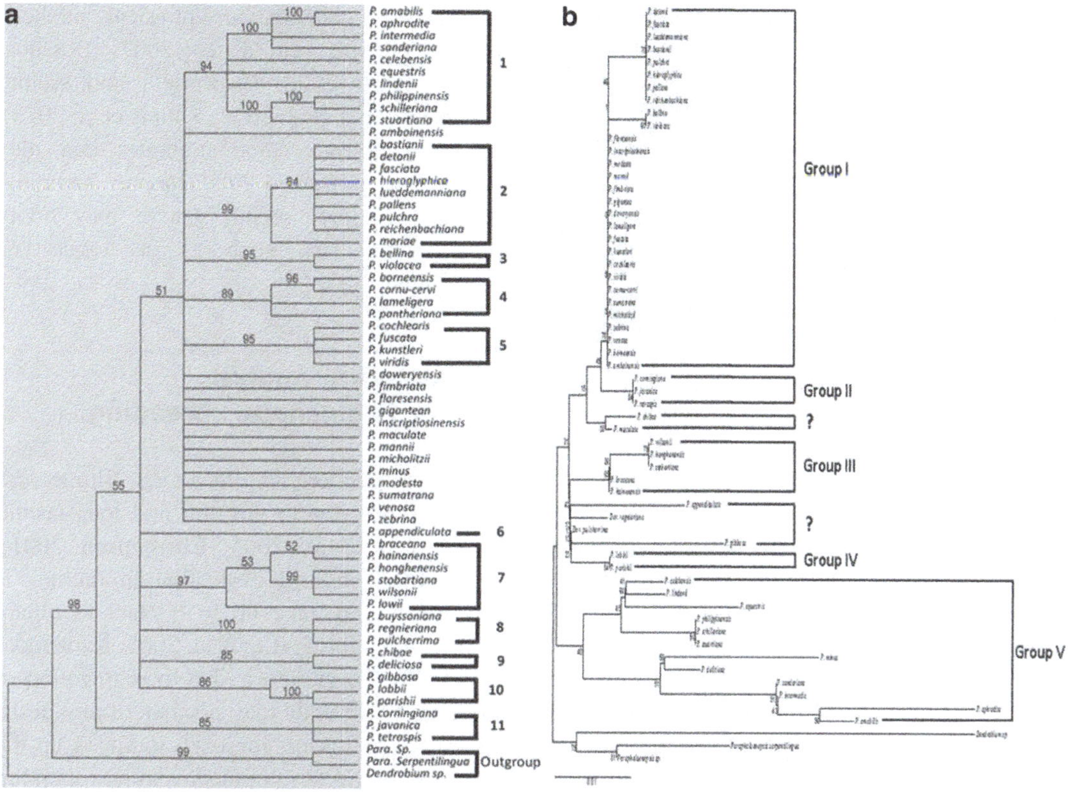

Fig. 2.1 Phylogenetic tree of species in *Phalaenopsis* based on analysis of chloroplast *matK* sequences, according to the neighbor-joining method. Numbers near branches are bootstrap values: **a** 10^6 replicates; **b** 10^3 replicates

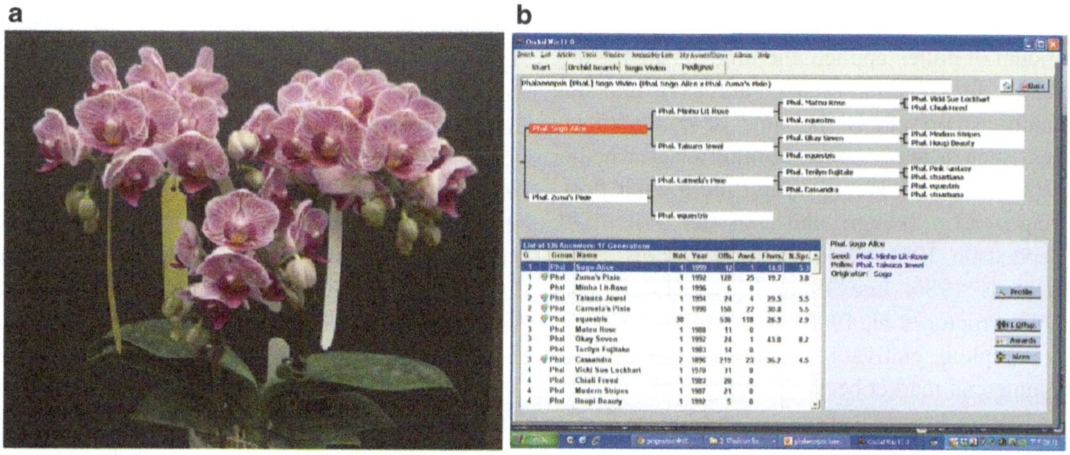

Fig. 2.2 **a** *Phalaenopsis* (*Doritaenopsis*) Sogo Vivien 'Sogo F858', a typical miniature hybrid cultivar. **b** The detailed genetic pedigree of F858 obtained with OrchidWiz, version 11.0

selfing, cross pollination and back cross (introgression) of any hybrid or species. The collection of pollen grains from desirable parents is especially important in a breeding program. However, during the progress of a breeding program, the desirable target parent may not be

Fig. 2.3 *Phalaenopsis* Liu's Twilight Rainbow 'Sogo F2006', a miniature cultivar that contains species *P. equestris*, *P. aphrodite* and *P. stuartiana* into its genetic constitution

available. We have studied the impact of freezing for the storage of *Phalaenopsis* pollinia, and our results suggest that pollinia may be stored for several years and still remain viable for cross pollination (Yuan et al. 2018).

Most commercial plants are propagated by seed; however, the reproductive success may be hindered by a series of factors, either intrinsic or extrinsic. Intrinsic factors may include the age of the flower selected for pollination, the variety or species, the ploidy and the presence of pollen incompatibility systems. Extrinsic factors may include the preservation condition of pollen, the amount of pollen applied, the time during the day chosen for pollination and the environmental conditions present at the time of conducting a cross such as temperature, humidity, and whether the flower is producing during the regular flowering season (Faegri and van der Pijl 1979; Stephenson 1981; De Vries and Dubois 1983; Richards 1986; Lee 1988; Lloyd and Schoen 1992; Proctor et al. 1996).

Polyploid cultivars are mostly pursued for commercial market because of their larger flower size and/or heavier substance, and longer flower life. Strategy for developing polyploidy includes the use of mitotic inhibitors such as colchicine to induce doubled chromosomes, exploiting unreduced gametes occurred naturally or by artificial induction by chemical or physical methods (Bolaños-Villegas and Chen 2007; Bolaños-Villegas et al. 2008; Brownfield and Köhler 2011; De Storme et al. 2012; Younis et al. 2014). Below we discuss some strategies that may contribute to advanced *Phalaenopsis* breeding. Breeding of other orchid genera may adopt similar approaches, such as in nobile-type *Dendrobium*.

2.3 Meiocyte Analysis for *Phalaenopsis* Breeding

For orchid breeders, *Phalaenopsis* Blume is a popular choice due to graceful and long-lasting flowers (Lin et al. 2005; Christenson 2001). Unfortunately, hybrid evaluation in orchids is slow and it may take up to 6 years to obtain blooming F2 hybrids (Lin et al. 2005; Kamemoto et al. 1999). Furthermore, due to segregation of characters F2 hybrids may not inherit any desirable traits at all. One approach to speed up the process of hybrid development is to pay attention to the process of segregation of chromosomes themselves; the logic behind this approach is that abnormalities in the patterns of segregation of chromosomes in the progenies may arise from physical differences in the structure and organization of the parental genomes. Ideally chromosomes should be observed under a light microscope with the uses of dyes such as carmine, orcein, fuchsin, or 4',6-diamidino-2-phenylindole (DAPI), a fluorescent DNA stain that intercalates between A/T nucleotide pairs and shines stably under UV light (Schwarzacher 2003). The structure and number of chromosomes is characteristic of each species and may vary because of evolutionary processes, such as inter-specific hybridization (Schubert 2007). During meiosis, related (e.g. homologous) chromosomes pair, recombine and exchange segments while in non-sexual tissue, chromosomes divide without recombination during mitosis (Wang and Copenhaver 2018).

2.4 Chromosome Behavior in *Phalaenopsis*

When chromosomes of *Phalaenopsis* show similar length and number, it is expected (but not guaranteed) that chromosomes from each parent may pair with their counterparts (Kao et al. 2001; Lin et al. 2005). In fact, the resulting progeny may inherit chromosomes that will pair and divide in a regular manner. However, if the parent species are not closely related, in most cases the progeny may inherit chromosomes with serious differences in length, and even number (aneuploidy) (Arends 1970; Lee et al. 2020). Hybrid chromosome assortments from divergent species may pair and divide irregularly during meiotic process (Lin et al. 2005). For instance, chromosomes may not pair at all (univalents), pair only partially, or pair in groups (multivalents), leading to losses of genetic material (Bolaños-Villegas et al. 2008). If chromosomes are lost, the lack of crucial DNA sequences may compromise basic metabolic processes, resulting in hybrids that fail to produce fertile pollen (Levin 2002; Dafni and Firmage 2002). Breeders may attempt to solve the problem by artificially doubling the chromosome number by arresting the cell division cycle with mitotic inhibitors such as colchicine or oryzalin (Hughdahl and Morejohn 1993). When each chromosome has a copy of itself, pairing can be restored, chromosomes divide almost normally and pollen fertility is increased (Kamemoto et al. 1999).

Unfortunately, chromosomes may undergo structural changes either spontaneously or under the influence of colchicine (Oliveira et al. 2004). A chromosome may invert or exchange segments with others (e.g. translocation, inversion, rearrangement) due to microhomology or non-homologous end joining (NHEJ) (Schmidt et al. 2019). Such rearrangements and accompanying deletion and amplification of DNA are major bottlenecks in the process of breeding new hybrids, and should they occur new rounds of chromosome doubling might not restore pairing and fertility but quite the opposite; also the frequency of mutations may increase due to positive feedback (Tanaka et al. 2002). This problem can be addressed through thorough examination of pollen before the respective donors are used in breeding programs. Moreover, reagents and methods to accomplish this are relatively inexpensive and simple.

2.5 Chromosome Analysis in Male Meiocytes

Observation of meiotic (e.g. sexual) chromosomes in pollen of *Phalaenopsis* is advantageous over use of mitotic tissue for several reasons, not only can the overall number of chromosomes be determined in a hybrid, but problems in pairing will be expressed as lagging 'orphan' metaphase chromosomes and pollen haploid daughter cells (e.g. tetrads) with irregular shape (Arends 1970). Unopen flower buds at 3/4 of the final size are believed to be most suitable for the observation of meiotic metaphase chromosomes (Dyer 1979). However, we have observed empirically that this pattern varies across hybrids, thus examination of each potential parent is advised. After collection in the greenhouse, immature pollinium can be directly placed on fixative solution. In order to soften samples, it may be advisable to resort to enzymes such as 0.5% (w/v) pectinase (Fluka, Buchs, Switzerland), and 1% (w/v) cellulase (Sigma-Aldrich, St. Louis) (Lin et al. 2005; Bolaños-Villegas et al. 2008). In case no such reagents are available, a bath in hot water at 60°C for one hour may help. Species that reproduce sexually reduce in half the chromosome number present in their gametes. This reduction is possible because of a specialized process of division commonly known as meiosis (Wang and Copenhaver 2018). Meiosis features two rounds of chromosome division (meiosis I and meiosis II) preceded by a single round of DNA replication just before meiosis I, and the end result are four haploid nuclei (Wang and Copenhaver 2018). These nuclei will then differentiate into gametes (Fayos et al. 2019). During meiosis II sister chromatids segregate, but in meiosis I it is the homologous chromosomes that segregate

(Wang and Copenhaver 2018). For cytological, genetic and molecular purposes, both phases of meiosis are further split into prophase, metaphase, anaphase and telophase, which are somehow similar to what is observed during mitosis (Wang and Copenhaver 2018). But prophase of meiosis I is unique and it is broken down into five stages: leptotene (formation of double-strand breaks), zygotene (pairing and synapsis), pachytene (synapsis is completed, strand breaks are fully repaired by recombination), diplotene (crossover sites become visible) and diakinesis (homologs separate except crossover sites are called chiasmata) (Wang and Copenhaver 2018).

Meiotic recombination that occurs during the prophase of meiosis I generates recombination products known as crossovers (Fayos et al. 2019). Crossovers result from a reciprocal exchange that takes place between two non-sister chromatids from two homologous chromosomes. Its cytological (visible) result is a chiasma (chiasmata in plural) (Fayos et al. 2019). The disruption of recombination and the failure to establish chiasmata between homologs results in defects in chromosome segregation and losses/gains in chromosome number (Wang and Copenhaver 2018; Fayos et al. 2019). Chromosomes that fail to segregate properly are eventually organized into small, independent nuclei unable to carry elemental cell processes (Ma 2006). In *Phalaenopsis* hybrids it is common to observe groups of tetrads with many micronuclei, these often show poor viability (Fig. 2.4), and frequent capsule abortion (Bolaños-Villegas et al. 2008). Conversely, an abundance of regular tetrads after meiosis may be considered as a good indicator of hybrid fitness, and vice versa (Fig. 2.5) (Bolaños-Villegas et al. 2008). How events in meiosis influence chromosome division and the formation of pollen daughter cells is clearly exemplified in Fig. 2.5, which shows a composite image of cells from different hybrids at the final stages of division. In *Phalaenopsis* unpaired chromosomes do not segregate in an orderly fashion or may not segregate at all. The presence of these laggards almost invariably leads to the formation of many abnormal daughter cells. In contrast, regular segregation and distribution of chromosomes during cell division precludes the formation of four daughter cells with the same morphology and size.

2.6 Pollen Viability

The relative ability of mature pollen to successfully develop a pollen tube (e.g. pollen viability) and fertilize an ovule is best studied in fresh, soft pollen exposed to dyes that indicate the presence of metabolic activity or cell membrane integrity, such as fluorescein diacetate (FDA) (Dafni and Firmage 2002). To do so, a small amount of pollen is excised and mildly ground over a glass slide, the dye is added, a coverslip is placed and the sample is incubated at room temperature in the dark. Then an actual count is done under an epifluorescence microscope. Typical viability values in *Phalaenopsis* may range from 20 to 70% (Bolaños-Villegas et al. 2008). As a note of caution, actual pollen viability is considered to vary over time, even under cold storage (Wang et al. 2004). Pollen viability is at best a simplification of complex phenomena affecting the capacity of pollen to extrude a pollen tube and transport two sperm nuclei into the embryo sac (Berger et al. 2008), thus the relationship between patterns of cell division during meiosis and pollen viability values may be misleading and need to be verified in vivo. Figure 2.4 shows

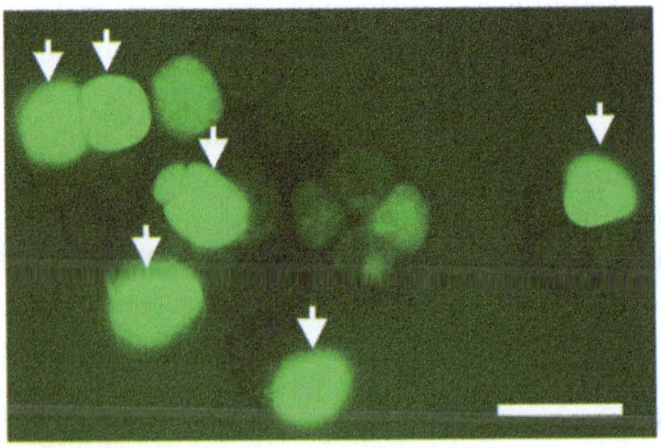

Fig. 2.4 Fluorescence microscope image of male tetrads in hybrid *Phalaenopsis* Ruey Lih Beauty after treatment with fluorescein diacetate (FDA) solution. Arrows point toward viable tetrads. Scale bar = 10 μm

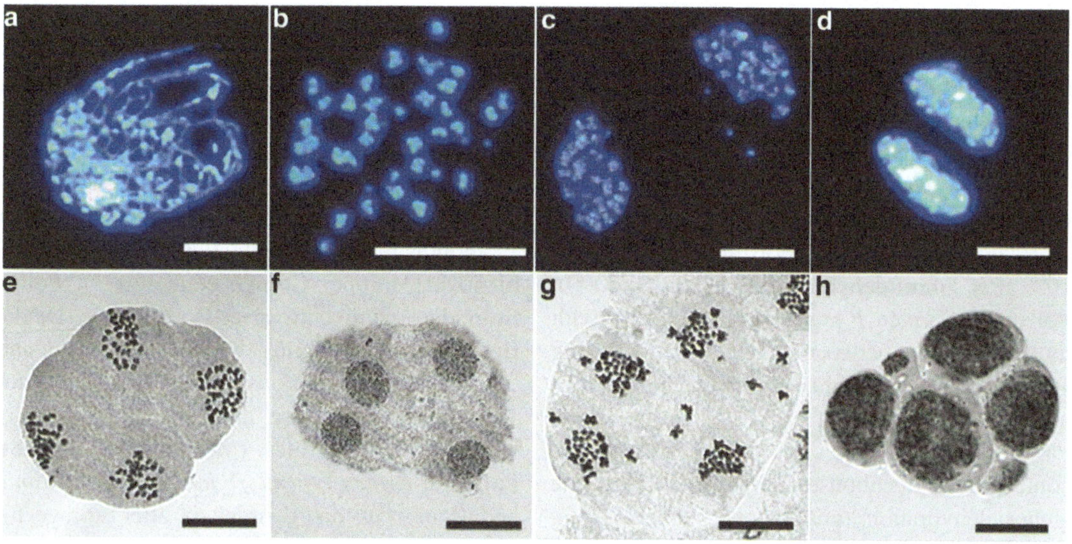

Fig. 2.5 Male meiosis in different *Phalaenopsis* hybrids. A-D, meiosis in *P.* Fuchsia Princess 'KHM648', as seen by DAPI staining. Pachytene (**a**), metaphase I (**b**), anaphase I (**c**), telophase I (**d**), telophase II, cell showing perfect chromosome segregation, *P. bellina* 'PI446' (**e**), cytokinesis, normal tetrad showing no micronuclei, *P.* Ben Yu Star 'Red Dragon' (**f**), anaphase II, cell showing laggards, *P.* Sogo Gem 'Sogo F859' (**g**), cytokinesis, abnormal tetrad with micronuclei, *P.* Chian Xen Diamond 'Sogo F1481' (**h**). Scale bar = 10 μm

male tetrads from hybrid *P.* Ruey Lih Beauty after treating with FDA solution. Viable tetrads exhibit fluorescence brightly after incubation.

2.7 Pollen Storage Management

Germplasm resource conservation refers to the collection and preservation of shoot meristems, seeds, zygotic and somatic embryos, to prevent loss of genetic diversity in species of interest (Kulus and Zalewska 2014; Ren et al. 2019; Wang et al. 2015; Yuan et al. 2018). Pollen is critical for the preservation, exchange and hybridization of germplasm resources (Masum-Akond et al. 2012; Ren et al. 2019). A successful breeding program requires the selection of superior parents for performing cross pollinations. However, elite cultivars may not be at hand always; therefore, pollen storage is an alternative to keep germplasm available around the clock (Hanna and Towill 1995; Sedgley and Harbard 1993; Wang et al. 2015). Developing a suitable strategy for pollen storage is significantly important, for instance long-term frozen storage. Several issues related to the pollen quality after storage need to be considered, including genetic background, water content and storage temperature. The latter two issues are discussed as follows.

1. **Water content**

The moisture content is an important factor affecting the storage of pollen for long-term preservation. Towill (1985) indicated that pollen could be classified into desiccation-tolerant or desiccation-sensitive. Based on the classification, some reports showed that pollen germination was significantly affected as drying temperature and time increased (Pacini et al. 2006; Wang et al. 2015; Yates et al. 1991). Yates et al. (1991) reported that the pollen of pecan showed a high germination rate when dried at 35 °C for 1 h. Pacini et al. (2006) further showed that pollen lost the germination viability followed by increased drying time. Wang et al. (2015) showed that pollen drying at 35 °C for 6 h is the most suitable method for preservation in the long term. In Paenoniaceae, the pollen retained high germination viability when dried at room temperature (23 ± 2 °C) for 24 h (Ren et al. 2019).

Barnabas et al. (1988) showed that maize pollen with its water content reduced at 13% had the highest germination rate. In *Dactylorhiza maculate*, pollen with a moisture content of 10% showed the highest germination rate, and pollen germination was inversely reduced by increasing the moisture content (Marks et al. 2014). Pollen of *Hylocereus* reached a moisture content of 5–10% after 1 h of dehydration (Metz et al. 2000). Pollen from seven *P. suffruticosa* cultivars with moisture content between 9 and 13% after drying at 105 °C for 2 h could still retain their germination viability (Ren et al. 2019). Furthermore, Parton et al. (2002) indicated that pollen moisture content and greenhouse humidity affected the anther dehydration time before storage. Hong et al. (1999) and Yates et al. (1991) proposed that among plant species the reduction of pollen water content to a certain level was advantageous for long-term preservation.

2. **Storage temperature**

Pollen longevity or pollen viability is defined as the retention of pollen germination ability after prolonged storage (Vaknin and Disikowitch 2000). Several reports have shown that pollen stored at low temperatures was effective for long-term preservation of plants, such as cherimoya (Lora et al. 2006), Arabidopsis (Bou Daher et al. 2009), bromeliads (Parton et al. 2002), caladium (Deng and Harbaugh 2004), jojoba (Vaknin et al. 2003), lily (Wang et al. 2004) and *Phalaenopsis* (Yuan et al. 2018). Our previous study on *Phalaenopsis* pollen storage demonstrated viable pollen stored at cold to frozen temperatures, including 4, −20, and −80 °C. Despite pollen could be stored at −196 °C for over four years, this will require sophisticated laboratory equipment (Wang et al. 2015). Alternatively, one can use a simple and cost-effective refrigerator to attain for long-term preservation. For instance, pollen from Litchi remains viable after 52 days of storage at 4 °C (Wang et al. 2015), and pollen from almond can be stored for 8 weeks at 4 °C (Martínez-Gómez et al. 2000). However, capacity to germinate after cold storage varies across species and cultivars. In seven species of Crape Myrtle (*Lagerstroemia* spp.), including *L. indica* 'Catawba', *L. fauriei* 'Kiowa', *L. limii*, *L. subcostata* and *L. speciosa*, pollen stored for 15 days at 4 °C retained germination rates from 10 to 35% (Masum-Akond et al. 2012). Pollen from Californian almond cultivars 'Nonpareil', 'Ne Plus Ultra' and 'Sonora', stored at 4 °C for 12 months retained germination at values from 8 to 50% (Martínez-Gómez et al. 2002). Pollen from date palm cultivars, 'Bouhlesse', 'Deglat Beida', 'Deglet Nour', 'Ghars', 'Halwaya'and 'Moch Deglet' that is stored at 4 °C for 13 months retains the ability to germinate at values from 19 to 66% (Mesnoua et al. 2018). Pollen of *Protea repens*, *P. magnifica*, *P. eximia* and *P. aristata* has germinated after storage for up to 270 days at 4 °C, and may show a 60% germination rate after 360 days (van der Walt and Littlejohn 1996). Yuan et al. (2018) showed that pollen of *P.* Little Gem Stripes stored at 4 °C for 40 weeks retained their germinability approximately 10–20%.

Some reports indicated that pollen remains its longevity when stored at sub-freezing temperature for some periods. Grape pollen retained more than 20% germination rate after 25 days of storage at both −18 and −40 °C (Tang et al. 2018). Indeed, Cohen et al. (1989) reported that grape pollen could be stored successfully at −18 °C for up to 6 months. Pollen from six date palm cultivars, belonging to 'Bouhlesse', 'Deglat Beida', 'Deglet Nour', 'Ghars', 'Halwaya' and 'Moch Deglet' that was stored at −20 °C for 13 months showed germination rates of 23–89% (Mesnoua et al. 2018). Interestingly, pollen from *Dactylorhiza fuchsii* that was stored for 6 years at −20 °C was still able to germinate at a 64% rate (Marks et al. 2014). Pritchard and Prendergast (1989) used several orchid species to evaluate the pollen storage ability, including *Anacamptis pyramidalis*, *Cymbidium elegans*, *C. tracyanum*, *Dactylorhiza fuchsii*, *D. maculata*, *Epipactis purpurata*, *Gymnadenia conopsea*, *Listera ovata*, *Orchis mascula*, *O. morio* and *Spiranthes spiralis*. The result indicated that pollen stored at both −20 and −196 °C may be preserved for at least one year. In our case we have observed that pollen from *Phalaenopsis*

Little Gem Stripes could be stored successfully at either −20 or −80 °C for up to 96 weeks (Fig. 2.6).

2.8 Determination of Pollen Viability After Storage

The evaluation of pollen viability can be performed by several chemical staining methods (Bellusci et al. 2010; Sorkheh et al. 2011; Abdelgadir et al. 2012). Hand pollination probably is the best and easiest way to estimate pollen viability after storage (Lyakh et al. 1998). It has been determined that storage at low temperature affects the ability to properly fertilize ovules, while pollen viability staining may still appear (Lyakh et al. 1998; Marks et al. 2014; El-Homosany and Sayed 2015). Metz et al. (2000) observed that pollen from *Hylocereus* that had been stored at 4 °C for 3–9 months showed fruit set rates of 60–70% after pollination, but storage at sub-freezing temperature led to 100% fruit set. However, when pollen from *Brassica napus* L. was stored at 3 or 10 °C, the resulting number of seeds after pollination decreased with longer storage periods (Lyakh et al. 1998). We obtained a 50% fruit set using pollen from *P.* Little Gem Stripes stored at both −20 and −80 °C for 96 weeks, which was placed on a receptive parent (Fig. 2.7). Mature capsules contained viable seeds. And in the case of *D. fuchsia*, pollen storage affected seed yield after pollination, but did not affect the germination ability of that seed (Marks et al. 2014).

2.9 Conclusions and Perspectives

A tremendous number of *Phalaenopsis* hybrids are created and registered with the Royal Horticultural Society (RHS, http://apps.rhs.org.uk/horticulturaldatabase/orchidregister/orchidregister.asp) by breeders from all over the world. The genetic background of most hybrids is rather complicated, so aspects such as cross incompatibility, hybrid sterility, and many morphological and physiological traits are ignored in favor of purely commercial aspects. Only in recent years there has been a shift to performing genomic and genetic research in order to understanding orchid growth and development (Huang et al. 2016). Through the application of research discussed in this book, in future it may be possible to perform more precise breeding toward the improvement of desirable traits. Methods that allow for sporad

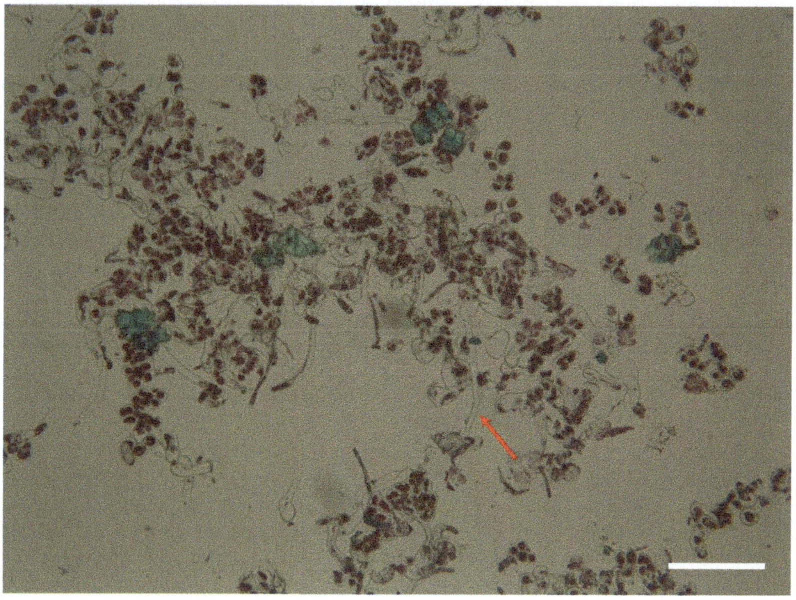

Fig. 2.6 Pollen tube growth in *Phalaenopsis* Little Gem Stripes as an indicator of viability after storage at −80 °C. The stored pollinia were germinated in vitro in Brewbaker and Kwack medium for 7 days and stained with Alexander Red dye. Arrow indicates pollen tube. Bars = 50 μm

Fig. 2.7 Capsules developed in the receptive parent *P*. Sogo Vivien 'Sogo F858' after pollination with pollinia stored at −20 or −80 °C. The hybrid seeds germinated readily after sowing in vitro

analysis (Bolaños-Villegas et al. 2008) can be used to evaluate formation of unreduced gametes after chemical and physical treatments, and may facilitate breeding of new polyploids. Newly discovered species may be added to the repertoire of breeders to create new interspecific hybrids with traits not currently observed in commercial hybrids. And collection and characterization of mutants derived from sexual hybridization or from somaclonal variation (Chen et al. 1998) may offer soon the opportunity to better understand orchid growth and development, and make it easier to control flowering (Duttke et al. 2012).

References

Abdelgadir HA, Johnson SD, Van Staden J (2012) Pollen viability, pollen germination and pollen tube growth in the biofuel seed crop *Jatropha curcas* (Euphorbiaceae). S Afr J Bot 79:132–139

Arends JC (1970) Cytological observations on genome homology in eight interspecific hybrids of *Phalaenopsis*. Genetica 41:88–100

Barnabas B, Kovacs G, Abranyi A, Pfahler P (1988) Effect of pollen storage by drying and deep-freezing in the expression of different agronomic traits in maize (*Zea mays* L.). Euphytica 39:221–225

Bellusci F, Musacchio A, Stabile R, Pellegrino G (2010) Differences in pollen viability in relation to different deceptive pollination strategies in Mediterranean orchids. Ann Bot 106:769–774

Berger F, Hamamura Y, Ingouff M, Higashiyama T (2008) Double fertilization: caught in the act. Trends Plant Sci 13:437–443

Bolaños-Villegas P, Chen FC (2007) Cytological identification of chromosomal rearrangements in *Doritaenopsis* and *Phalaenopsis*. J Int Coop 2:1–11

Bolaños-Villegas P, Chin SW, Chen FC (2008) Meiotic chromosome behavior and capsule setting in *Doritaenopsis* hybrids. J Amer Soc Hortic Sci 133:107–116

Bou Daher F, Chebli Y, Geitmann A (2009) Optimization of conditions for germination of cold-stored *Arabidopsis thaliana* pollen. Plant Cell Rep 28:347–357

Brownfield L, Köhler C (2011) Unreduced gamete formation in plants: mechanisms and prospects. J Exp Bot 62:1659–1668

Chen WH, Chen TM, Fu YM, Hsieh RM, Chen WS (1998) Studies on somaclonal variation in *Phalaenopsis*. Plant Cell Rep 18:7–13

Christenson EA (2001) Phalaenopsis-A Monograph. Timber Press, Portland, Oregon, p 330

Cohen E, Lavi U, Spiegel-Roy P (1989) Papaya pollen viability and storage. Sci Hortic 40:317–324

Dafni A, Firmage D (2002) Pollen viability and longevity: practical, ecological, and evolutionary implications. Plant Syst Evol 222:113–132

Deng Z, Harbaugh BK (2004) Technique for in vitro pollen germination and short-term pollen storage in caladium. HortScience 39:365–367

De Storme N, Copenhaver GP, Geelen D (2012) Production of diploid male gametes in Arabidopsis by cold-induced destabilization of postmeiotic radial microtubule arrays. Plant Physiol 160:1808–1826

De Vries DP, Dubois LAM (1983) Pollen and pollination experiments. X. The effect of repeated pollination on fruit and seed-set in crosses between the hybrid tea-rose CVS. Sonia and Ilona. Euphytica 32:685–689

Duttke S, Zoulias N, Kim M (2012) Mutant flower morphologies in the wind orchid, a novel orchid model species. Plant Physiol 158:1542–1547

Dyer AF (1979) Investigating Chromosomes. Edward Arnold Pub, London, p 430

El-Homosany AA, Sayed HA (2015) Effect of low temperature and cryopreservation on in vitro pollen germination of some olive cultivars. Am-Euras. J Agric Environ Sci 15:1803–1808

Faegri K, van der Pijl L (1979) The Principles of Pollination Ecology, 3rd edn. Pergamon Press, Oxford, England, p 244

Fayos I, Mieulet D, Petit J, Meunier AC, Perin C, Nicolas A, Guiderdoni E (2019) Engineering meiotic recombination pathways in rice. Plant Biotechnol J 17:1–16

Hanna WW, Towill LE (1995) Long-term pollen storage. Plant Breed Rev 13:179–199

Hong TD, Ellis RH, Buitink J, Walters C, Hoekstra FA, Crane J (1999) A model of the effect of temperature and moisture on pollen longevity in air-dry storage environments. Ann Bot 83:167–173

Huang JZ, Lin CP, Cheng TC, Huang YW, Tsai YJ, Cheng SY, Chen YW, Lee CP, Chung WC,

Chang BCH, Chin SW, Lee CY, Chen FC (2016) The genome and transcriptome of *Phalaenopsis* yield insights into floral organ development and flowering regulation. PeerJ 4:e2017

Hughdahl JD, Morejohn LC (1993) Rapid and reversible high-affinity binding to dinitroaniline herbicide oryzalin to tubulin from *Zea mays* L. Plant Physiol 102:725–740

Kamemoto H, Amore DT, Kuehnle A (1999) Breeding Dendrobium Orchids in Hawaii. University of Hawaii Press, Honolulu, p 166

Kao YY, Chang SB, Lin TY, Hsieh CH, Chen YH, Chen WH, Chen CC (2001) Differential accumulation of heterochromatin as a cause for karyotype variation in *Phalaenopsis* orchids. Ann Bot 87:387–395

Kulus D, Zalewska M (2014) Cryopreservation as a tool used in long-term storage of ornamental species- A review. Sci Hortic 168:88–107

Lee TD (1988) Patterns of Fruit and Seed Production. In: Doust JL, Doust LL (eds) Plant Reproductive Ecology—Patterns and Strategies. Oxford University Press, Oxford, pp 179–202

Lee YI, Tseng YF, Lee YC, Chung MC (2020) Chromosome constitution and nuclear DNA content of *Phalaenopsis* hybrids. Sci Hortic 262:109089

Levin DA (2002) The Role of Chromosomal Change in Plant Evolution. Oxford Univ Press, New York, p 230

Lin CC, Chen YH, Chen WH, Chen CC, Kao YY (2005) Genome organization and relationships of *Phalaenopsis* orchids inferred from genomic in situ hybridization. Bot Bull Academia Sinica 46:339–345

Lin TP (1977) Native Orchids of Taiwan, Southern Materials Ctr., Taipei, pp 353

Lloyd DG, Schoen DJ (1992) Self- and cross-fertilization in plants. I Functional dimensions. Int J Plant Sci 153:358–369

Lora J, Pérez de Oteyza MA, Fuentetaja P, Hormaza JI (2006) Low temperature storage and in vitro germination of cherimoya (*Annona cherimola* Mill.) pollen. Sci Hortic 108:91–94

Lyakh VA, Soroka AI, Kalinova MG (1998) Pollen storage at low temperature as a procedure for the improvement of cold tolerance in spring rape, *Brassica napus* L. Plant Breed 117:389–391

Ma H (2006) A molecular portrait of Arabidopsis meiosis. In: Somerville CR and Meyerowitz EM (eds). The Arabidopsis Book 4: e0095

Marks TR, Seaton PT, Pritchard HW (2014) Desiccation tolerance, longevity, and seed-siring ability of entomophilous pollen from UK native orchid species. Ann Bot 114:561–569

Martínez-Gómez P, Gradziel TM, Ortega E, Dicenta F (2000) Short-term storage of almond pollen. HortScience 35:1151–1152

Martínez-Gómez P, Gradziel TM, Ortega E, Dicenta F (2002) Low temperature storage of almond pollen. HortScience 37:691–692

Masum-Akond ASMG, Pounders CT, Blythe EK, Wang XW (2012) Longevity of Crape Myrtle pollen stored at different temperatures. Sci Hortic 139:53–57

Mesnoua M, Roumani M, Salem A (2018) The effect of pollen storage temperatures on pollen viability, fruit set and fruit quality of six date palm cultivars. Sci Hortic 236:279–283

Metz C, Nerd A, Mizrahi Y (2000) Viability of pollen of two fruit crop cacti of the genus *Hylocereus* is affected by temperature and duration of storage. HortScience 35:22–24

Oliveira VM, Forni-Martins ER, Magalhães PM, Alves MN (2004) Chromosomal and morphological studies of diploid and polyploid cytotypes of *Stevia rebaudiana* (Bertoni) Bertoni (Eupatorieae, Asteraceae). Genet Mol Biol 27:215–222

Pacini E, Guarnieri M, Nepi M (2006) Pollen carbohydrates and water content during development, presentation, and dispersal: a short review. Protoplasma 228:73–77

Parton E, Vervaeke I, Delen R, Vandenbussche B, Deroose R, De Proft M (2002) Viability and storage of bromeliad pollen. Euphytica 125:155–161

Pritchard HW, Prendergast FG (1989) Factors influencing the germination and storage characteristics of orchid pollen. In: Pritchard HW (ed) Modern Methods in Orchid Conservation: the Role of Physiology, Ecology, and Management. Cambridge Univ Press, England, pp 1–16

Proctor M, Yeo P, Lack A (1996) The Natural History of Pollination. Timber Press, Portland, Oregon, p 479

Ren R, Li Z, Li B, Xu J, Jiang X, Liu Y, Zhang K (2019) Changes of pollen viability of ornamental plants after long-term preservation in a cryopreservation pollen bank. Cryobiology 89:14–20

Richards AJ (1986) Plant Breeding Systems. George Allen and Unwin, London, p 529

Sedgley M, Harbard J (1993) Pollen storage and breeding system in relation to controlled pollination of four species of *Acacia* (Leguminosae: Mimosoideae). Aust J Bot 41:601–609

Schmidt C, Schindele P, Puchta H (2019) From gene editing to genome engineering: restructuring plant chromosomes via CRISPR/Cas. aBIOTECH 1:21–31

Schubert I (2007) Chromosome evolution. Curr Opin Plant Biol 10:109–115

Schwarzacher T (2003) DNA, chromosomes, and in situ hybridization. Genome 46:953–962

Sorkheh K, Shiran B, Rouhi V, Khodambashi M (2011) Influence of temperature on the in vitro pollen germination and pollen tube growth of various native Iranian almonds (*Prunus* L. spp.) species. Trees 25:809–822

Stephenson AG (1981) Flower and fruit abortion: proximate causes and ultimate functions. Annu Rev Ecol Evol Syst 12:253–279

Tang SS, Huang YL, Fu X (2018) A comparative study on the pollen viability and storage conditions of the nuclear and non-nuclear varieties. Trends Hortic 15–18

Tanaka H, Tapscott SJ, Trask BJ, Yao MC (2002) Short inverted repeats initiate gene amplification through the formation of a large DNA palindrome in mammalian cells. Proc Natl Acad Sci USA 99:8772–8777

Towill LE (1985) Low temperature and freeze-/vacuum-drying preservation of pollen. In: Kartha KK (ed) Cryopreservation of Plant Cells and Organs. CRC Press, Boca Raton, pp 171–198

Tsai CC, Huang SC, Chou CH (2005) Molecular phylogeny of *Phalaenopsis* Blume (Orchidaceae) based on the internal transcribed spacer of the nuclear ribosomal DNA. Plant Syst Evol 256:1–16

Tsai CC, Chiang YC, Huang SC, Chen CH, Chou CH (2010) Molecular phylogeny of *Phalaenopsis* Blume (Orchidaceae) on the basis of plastid and nuclear DNA. Plant Syst Evol 228:77–98

Tsao JY (2003) Phylogenetic Analysis of *Phalaenopsis* Species Based on Plastid matK and rbcL Sequences. Master Thesis, Department of Tropical Agriculture and International Cooperation, National Pingtung Univ Sci &Technol, Taiwan. 111 pp

Younis A, Hwang YJ, Lim KB (2014) Exploitation of induced 2n-gametes for plant breeding. Plant Cell Rep 33:215–223

Vaknin Y, Disikowitch D (2000) Effects of short-term storage on germinability of pistachio pollen. Plant Breed 119:347–350

Vaknin Y, Mills D, Benzioni A (2003) Pollen production and pollen viability in male jojoba plants. Indust Crops Product 18:117–123

van der Walt ID, Littlejohn GM (1996) Storage and viability testing of *Protea* pollen. J Amer Soc Hortic Sci 121:804–809

Wang LM, Wu JF, Chen JZ, Fu DW, Zhang CY, Cai CH, Ou LG (2015) A simple pollen collection, dehydration, and long-term storage method for litchi (*Litchi chinensis* Sonn.). Sci Hortic 188:78–83

Wang ML, Hsu CM, Chang LC, Wang CS, Su TH, Huang YJJ, Jiang L, Jauh GY (2004) Gene expression profiles of cold-stored and fresh pollen to investigate pollen germination and growth. Plant Cell Physiol 45:1519–1528

Wang Y, Copenhaver GP (2018) Meiotic recombination: mixing it up in plants. Annu Rev Plant Biol 69:577–609

Yates IE, Sparks D, Connor K, Towill L (1991) Reduced pollen moisture simplifies long-term storage of pecan pollen. J Amer Soc Hortic Sci 116:430–434

Yuan SC, Chin SW, Lee CY, Chen FC (2018) *Phalaenopsis* pollinia storage at sub-zero temperature and its pollen viability assessment. Bot Stud 59:1

Yukawa T, Kita K, Handa T, Hidayat T, Ito M (2005) Molecular phylogenetics of *Phalaenopsis* (Orchidaceae) and allied genera: Re-evaluation of generic concepts. Acta Phytotaxonomica et Geobotanica 56:141–161

The Tiny Twig Epiphyte *Erycina pusilla*, a Model for Orchid Genome and Breeding Research

3

Pablo Bolaños-Villegas, Chen Chang, and Fure-Chyi Chen

Abstract

Erycina pusilla is an orchid species from the tropics with a wide distribution in neotropical America. It is a fast-growing tiny twig epiphytic orchid and blooms constantly throughout the year. Approximately 28 MIKC/MADS-box genes from *E. pusilla* have been identified. Type-II MADS-box proteins (also called MIKC-domain proteins) play a crucial role during flower development, making this species an ideal orchid model for developmental studies. *E. pusilla* has a small diploid genome and two different alternative cytotypes have been reported, with 10 and 12 chromosomes. The karyotype evolution in *Erycina* is the result of processes of segment translocation and heterochromatin expansion/deletion. In order to breed new commercial orchid species, *E. pusilla* has been crossed with several important Oncidiinae orchids. The clone PSYP1 as *E. pusilla* 'Hsingda Golden' derived from in vitro flowering system has been granted the Plant Variety Rights in Taiwan for protection. The *E. pusilla* plants can also be engineered using the *Agrobacterium tumefaciens* strain EHA105 together with a plasmid vector carrying the CaMV 35S promoter system. The twig epiphyte *E. pusilla* holds enormous biotechnological and horticultural potential. It may be necessary to attempt more basic and applied research to transform the current knowledge into new *Oncidium* varieties with altered flower morphology, enhanced growth rate and vibrant colors.

3.1 Introduction

Orchids are one of the most diverse and numerous angiosperm families. Because of the complexity of their biological features, orchids are considered as good testing systems to check test on fundamental hypotheses. It is believed that speciation is high in orchids, suggesting that evolutionary processes are quite active in these plants (Duttke et al. 2012; Tsai et al. 2017).

Within monocots the orchid clade represents a petaloid group that is different from other model systems such as maize (*Zea mays*), and

P. Bolaños-Villegas (✉)
Fabio Baudrit Agricultural Research Station, University of Costa Rica, La Garita, Alajuela 20101, Costa Rica
e-mail: pablo.bolanosvillegas@ucr.ac.cr

P. Bolaños-Villegas
Jardín Botánico Lankester, Universidad de Costa Rica, Cartago P.O. Box 302-7050 Costa Rica

C. Chang (✉)
Department of Horticulture, National Chung Hsin University, Taichung, Taiwan
e-mail: chenchang@dragon.nchu.edu.tw

F.-C. Chen
Department of Plant Industry, National Pingtung University of Science & Technology, Pingtung, Taiwan

F.-C. Chen and S.-W. Chin (eds.), *The Orchid Genome*, Compendium of Plant Genomes,
https://doi.org/10.1007/978-3-030-66826-6_3

Arabidopsis thaliana (Duttke et al. 2012). Other unique characteristics are the presence of symbiotic relationships between orchids and mycorrhizal fungi (Dearnaley et al. 2012), photosynthesis based on either C3 or crassulacean acid metabolism (CAM) (Tsai et al. 2017), and most growth as epiphytes (Heyduk et al. 2019). Because of their successful adaptation to the environment, orchids have evolved unique reproductive strategies (Tsai et al. 2017). However, the study of orchid flower development from a molecular and genetic perspective is still in its infancy (Duttke et al. 2012), mostly because it is necessary to test hypotheses with a model species that is amenable to growth in the laboratory, has assortment of mutants to choose from (Duttke et al. 2012) and allows for gain-of-function studies (Pan et al. 2012).

Twig epiphytes are a small group of orchids which is restricted to the smallest and thinnest branches of their host plants. These branches receive high doses of sunlight radiation, the atmospheric humidity surrounding them is low and there is a low accumulation of minerals. This habitat fragility demands that its inhabitants rapidly reach reproductive age in order to produce seedlings that colonize new branches and to persist in such a dynamic environment (Dodson 1957; Chase 1986; Chase and Palmer 1997; Chase et al. 2005; Mondragón et al. 2007). For instance, in Chiapas state in Mexico it was reported that the twig epiphyte *Erycina crista-galli* reaches sexual maturity at the age of one year with the minimum plant size when compared to other twig epiphytes such as *Artorima erubescens*, *Laelia speciosa*, *Lepanthes caritensis* and *Lepanthes eltoroensis* (Mondragón et al. 2007). It is believed that the reproductive precocity of *E. crista-galli* is the result of the reduction of organs and the fusion of functions (photosynthetic roots, foliar trichomes with absorption capacity) which contribute to the optimization of physiological processes (Mondragón et al. 2007).

Erycina pusilla is well distributed in the American tropics (Baker and Baker 2006). It is an epiphyte that grows very quickly, has few chromosomes (a haploid number of only 6) and a very small genome 1.5 pg/1C (Félix and Guerra 2000; Chase et al. 2005). Plants develop deceptive flowers that cannot self-pollinate but are self-compatible. It is believed that oil-collecting bees are its pollinators (Dirks-Mulder et al. 2017). New tissue culture techniques and quick flowering mean that *E. pusilla* (cv. Hsingda Golden) will produce flowers (Fig. 3.1) and eventually set fruit in vitro within one year (Chiu et al. 2011a, b; Chiu and Chang 2018).

The plants are small (4–8 cm in height) and bloom constantly throughout the year (Sheehan and Sheehan 1989; Félix and Guerra 2000; Baker and Baker 2006; Chiu and Chang 2011a, b). It is also possible to mutagenize seedlings (Lin et al. 2013) and perform transformation with *Agrobacterium tumefaciens* (Lee et al. 2015). These traits make *E. pusilla* an attractive model for studies in flowering and functional genomics, and also a good parent for traditional experiments on hybridization (Liu et al. 2011; Chiu and Chang 2012, 2018).

3.2 MADS-Box Genes

A stereotypical orchid flower has a structural array of four whorls of organs. Results in the model plant *Arabidopsis thaliana* indicate that genes expressed in petals of *A. thaliana* are also expressed in the first whorl of petaloid monocots such as orchids (Dirks-Mulder et al. 2017). Similarity in the architecture of sepals and petals across lilies, gingers and orchids is believed to be caused by the consistent expression of such genes in tissues of the first whorl (Dirks-Mulder et al. 2017). It is believed that the regulation of floral development is carried by transcription factors from the MADS-box family (Bowman et al. 1989; Lin et al. 2016). The corresponding MADS-box proteins featured high conservation of an amino-terminal sequence of 60 residues called the MADS-box, or the M-domain (Lin et al. 2016).

Within MADS-box proteins, there is group called the Type-II, or known as MIKC-type

Fig. 3.1 The plant morphology (**a**) and in vitro flowering of *Erycina pusilla* (**b**) (bar = 1 cm)

proteins because of the feature of highly conserved domains, such as the MADS-domain (M), an intervening (I) domain, a keratin-like (K) domain, plus the highly variable carboxy terminal (C) domain (Álvarez-Buylla et al. 2000). The MIKC-type genes have been further split into two types, the MIKC^{C} and the MIKC, based on the divergence of sequences within and downstream of the I-domain (Álvarez-Buylla et al. 2000). The functional diversification of MIKC^{C} genes may account for the existence of differences in development and morphology in floral organs (Álvarez-Buylla et al. 2000), which are often explained through the ABCDE model of floral development (Theißen 2001). In this model the overlap of gene expression in stem cells of the meristem determines organ identity: sepals feature expression of A and E genes, petals show expression of A, B, and E, stamens of B, C, and E, carpels of C and E and ovules of D and E (Ditta et al. 2004). The representative genes are *APETALA* (*AP1*) for the A-class, *APETALA3* (*AP3*) and *PISTILLATA* (*PI*) for the B-class, *AGAMOUS* (*AG*) for the C-class, *SEEDSTICK* (*STK*) for the D-class, and *SEPALLATA* (*SEP*) for the E class (Dirks-Mulder et al. 2017). For *Oncidium* an alternative model called the perianth code has been proposed, in which duplicated genes from the B class (*AP3*) and E class (*AGL6*) compete among themselves to specify the formation of lip and petals, correspondingly (Gravendeel and Dirks-Mulder 2015).

The MADS-box *AGL6* is in fact one of the most notable genes in *Oncidium* and in *Phalaenopsis* and has been shown to regulate lip formation (Hsu et al. 2015; Huang et al. 2016). The *Phalaenopsis PhAGL6b* mutants are big lip mutants that show a 40–70% reduction in its expression as well as alternative splicing (Huang et al. 2016). In *E. pusilla* three different copies of *AGL6* have been identified. *EpMADS3* shows high expression in sepals and petals, *EpMADS4* shows high expression in lateral sepals, whereas *EpMADS5* is expressed highly in the lip and callus (Dirks-Mulder et al. 2017). In total approximately 28 MIKC/MADS-box genes have been identified in *E. pusilla*, making this species a suitable model for developmental studies in orchids (Lin et al. 2016; Dirks-Mulder et al. 2017). Using data from transcriptional studies and sequencing of BACs, it was possible to determine the corresponding cDNA sequences in full length and the respective distribution of introns and exons positions for these prospective 28 *EpMADS* genes (Lin et al. 2016). Unlike in model plant *Arabidopsis*, most MADS-box genes from *E. pusilla* feature introns longer than ten kilobases (Lin et al. 2016). Notably, *E. pusilla* lacks the famous MADS-box gene *FLOWERING LOCUS C* (*FLC*), a floral repressor found in *Arabidopsis* that is believed to be regulated by vernalization (Deng et al. 2011). In *E. pusilla* the optimal temperature for flowering is 27 °C, suggesting that vernalization is not part of its life cycle (Lin et al. 2016).

3.3 Genetic Transformation

Several orchid genera, such as *Phalaenopsis* and *Oncidium* are economically high value ornamental crops in places such as Taiwan (Lee et al. 2015). Therefore, effective transformation technologies have been developed to advance functional genetic studies in these orchids (Lee et al. 2015). It has been reported that *E. pusilla* plants can be successfully transformed using the *A. tumefaciens* strain EHA105 carrying a binary plasmid. For the plasmid the promoter of choice is the CaMV *35S* sequence (Lee et al. 2015). Explants can be selected on hygromycin-containing medium added with the growth regulators 6-benzylaminopurine and α-naphthaleneacetic acid for up to five months. Experiments showed that 3-month-old protocorm-like bodies (PLBs) represent the optimal stage for transformation. Self-pollination allowed the obtention of T_1 progenies in only 18 months (Lee et al. 2015; Li and Chan 2018).

In order to induce protocorm formation and proliferation, self-pollinated seed capsules of *Erycina pusilla* are dissected aseptically (Lapjit and Tseng 2015). Seeds are sterilized with a solution of NaOCl (1% commercial chlorine), vortexed by 15 min, and then washed with sterile water 3–5 times (Lapjit and Tseng 2015). The seeds are germinated in plastic dishes containing sterile half strength MS medium (2.2 g/L of Murashige and Skoog salts, 30 g/L of sucrose, 8 g/L of agar, at a pH of 5.7) (Lapjit and Tseng 2015). The Petri dishes are sealed with Parafilm and incubated inside a growth chamber at 25°C with a 16-hour light photoperiod (Lapjit and Tseng 2015). After 2 months of germination, germinating plantlets are transferred from Petri dishes to glass jars containing ½ MS medium (Lapjit and Tseng 2015). Three months after germination, protocorms should be 1 cm in length and green (Li and Chan 2018).

It may be very tempting to transform *Erycina* with CRISPR/Cas9 constructs in order to edit out MADS-box genes and shape the morphology of flowers. *Agrobacterium*-mediated RNA interference has been unsuccessfully attempted in the past (Lin 2012).

3.4 Compatibility for Breeding with Alliance Species

For the commercial breeding of new hybrids, *E. pusilla* has been crossed with other Oncidiinae from genera *Rodriguezia* and *Tolumnia* (Liu et al. 2011). Conclusive evidence of phylogenetic relatedness was obtained by comparing five chloroplast sequences across commercially relevant genera *Ada, Beallara, Comparettia, Degarmoara, Erycina, Huangara, Ionocidium, Ionopsis, Macradenia, Miltassia, Odontocidium, Odontoglossum, Oncidesa, Oncidium, Rodriguezia, Tolumnia, Zelemnia* and *Zelenkocidium* (Pan et al. 2012), with the chloroplast sequences *trnH-psbA, matK, trnF-ndhJ* and *IRb-SSC* (Pan et al. 2012). Sequence analysis showed that *E. pusilla*, *Rodriguezia* and *Tolumnia* are clustered together and are evolutionarily separate from genera *Oncidium*, *Odontocidium* and *Beallara* (Liu et al. 2011; Chiu and Chang 2012; Pan et al. 2012).

Efficient polyploidization has been reported with *Erycina* seedlings in tissue culture by exposing them to 0.5 to 1% colchicine for 4 days. Various mitotic chromosome numbers were observed in plants recovered, such as $2n = 2x = 12$, $2n = 4x = 24$, and $2n = 43$ (Lin et al. 2013).

According to results from Chiu and Chang (2018), it is possible to produce flasks of blooming ornamental *E. pusilla* plants one year after seed germination in vitro (Fig. 3.2). This cultivation method can also be used for the selection of desirable traits in plants that are the result of cross hybridization. Chiu and Chang (2018) named their *E. pusilla* clone PSYP-1 as ‘Hsingda Golden’, and then filed an application for plant variety rights in Taiwan in September 2013.

Fig. 3.2 An in vitro flowering plant of *E. pusilla* 'Hsingda Golden' grown for ornamental purpose

3.5 Genome and Chromosome Organization

Previous reports indicate that *Erycina pusilla* has a small diploid genome of approximately 1.475 Gb (Dirks-Mulder et al. 2017). Two alternate chromosome numbers have been reported, one of 2n = 10, and another of 2n = 12 (Yeh et al. 2015; Lin et al. 2013), but it is possible that the difference is caused by comparisons between different species (Yeh et al. 2017). Most notably the divergent chromosome count of 12 corresponds to a South American individual from Surinam (Yeh et al. 2017), whose flowers are different from those observed by us in Costa Rica at the Lankester Garden. *In situ* hybridization studies show that in mitotic chromosomes belonging to the 2 = 10 cytotype, there is a pair of 45S signals, and three distinguishable pairs of 5S rDNA signals. The signals from the 45S rDNA genes co-localize with 5S rDNA signals at the terminus of each long chromosome arm (Yeh et al. 2017) while telomeric foci appear on the terminus of each chromosome. However, in the chromosomes from the presumed cytotype from Surinam, there are four telomeric signals embedded in centromeres (Yeh et al. 2017). In general, the regular cytotype shows asymmetric chromosomes with pericentromeric heterochromatin, and does not show satellite chromosomes or nucleolar-organizing regions (NORs). It is believed that the ancestral chromosome number is 14, that 12 and 10 are more derived, and that karyotype evolution in *Erycina* is the result of processes of segment translocation and heterochromatin expansion/deletion (Yeh et al. 2017). It may also be tempting to probe whether the genome has evolved in order to cope with the brutal regime of radiation and desiccation experimented by twig epiphytes (Melters et al. 2012).

3.6 Metabolism and Physiology

In species with C3 metabolism, stomata stay open during the day in order to fixate CO_2, but water deficit may force stomata to shut off, resulting in truncated CO_2 fixation and reduced growth (Heyduk et al. 2019). Nonetheless, as a response to water stress, plants have evolved as a mechanism for concentrating CO_2 called Crassulacean acid metabolism (CAM) (Heyduk et al. 2019). In this mechanism CO_2 is concentrated by RuBisCO during the day, thus increasing its efficiency, while during the night respiration becomes the dominant reaction (Cai et al. 2015). Many orchids show preferential adoption of the CAM photosynthesis over the C3 pathway, perhaps as an adaptation to their epiphytic lifestyle that limits water supply (Cai et al. 2015). In *Phalaenopsis equestris* the carbonic anhydrase (CA) gene has undergone six duplications when compared to *Arabidopsis thaliana*, and this is thought to be highly significant trait since the CA enzyme catalyzes the reaction that converts CO_2 into carbonate, the very first step in CO_2 fixation (Cai et al. 2015). Time-course RNA expression studies suggest that *Erycina pusilla* shows an increase in the expression of key enzyme PEPC, which regulates the second step of the CAM pathway: fixation of HCO_3 in the chloroplast using phosphoenolpyruvate (PEP) as the substrate in order to produce oxaloacetate (OAA), the precursor of malic acid, which is later stored in the vacuole (Heyduk et al. 2019). Without histological evidence it is hard to pinpoint what are the exact metabolic and physiological adaptations undergone by *Erycina*.

3.7 Conclusions

The twig epiphyte *Erycina pusilla* holds enormous biotechnological and horticultural potential. The in vitro flowering and cross hybridization have been achieved by one of the author's lab (CC) for some commercial applications. It may be necessary to attempt more basic and applied research combined with the transcriptomic and genomic data in order to transform the current knowledge into new *Oncidium* varieties with altered flower morphology, enhanced growth rate and vibrant colors, and maybe other traits such as fragrance.

References

Álvarez-Buylla ER, Pélaz S, Liljegren SJ, Gold SE, Burgeff C, Ditta GS, Ribas de Pouplana L, Martínez-Castilla L, Yanofsky MF (2000) An ancestral MADS-box gene duplication occurred before the divergence of plants and animals. Proc Natl Acad Sci USA 97:5328–5333

Baker ML, Baker CO (2006) Orchid Species Culture *Oncidium/Odontoglossum* Alliance. 992p, Timber Press, Oregon

Bowman JL, Smyth DR, Meyerowitz EM (1989) Genes directing flower development in Arabidopsis. Plant Cell 1:37–52

Cai J et al (2015) The genome sequence of the orchid *Phalaenopsis equestris*. Nature Genet 47:65–72

Chase MW (1986) A reappraisal of the oncidioid orchids. Sys Bot 11:477–491

Chase MW, Hanson L, Albert VA, Whitten WM, Williams NH (2005) Life history evolution and genome size in subtribe Oncidiinae (Orchidaceae). Ann Bot 95:191–199

Chase MW, Palmer JD (1997) Leapfrog radiation in floral and vegetative traits among twig epiphytes in the orchid subtribe Oncidiinae. In: Giunish TJ, Sytsma KJ (eds) Molecular Evolution and Adaptive Radiation. Cambridge Univ. Press, Cambridge, pp 331–352

Chiu YT, Chang C (2011) A survey of biology research in *Erycina pusilla*. Seed & Nursery (Taiwan) 13:21–30

Chiu YT, Lin CS, Chang C (2011) In vitro fruiting and seed production in *Erycina pusilla*. Propagation Ornamental Plants 11(3):131–136

Chiu YT, Chang C (2012) Intergeneric crosses between *Erycina pusilla* and *Tolumnia* cultivars. J Taiwan Soc Hortic Sci 58:117–124

Chiu YT, Chang C (2018) In vitro flowering and breeding of *Erycina pusilla*. In: Lee YI, Yeung ECT (eds) Orchid Propagation: From Laboratories to Greenhouses-Methods and Protocols. Springer-Verlag, Switzerland AG, pp 257–265

Dearnaley JDW, Martos F, Selosse MA (2012) Orchid mycorrhizas: molecular ecology, physiology, evolution, and conservation aspects. In: Fungal Associations, Hock, B. (ed), Springer-Verlag, Switzerland AG, pp 207–230

Deng W, Ying H, Helliwell CA, Taylor JM, Peacock WJ, Dennis ES (2011) FLOWERING LOCUS C (FLC) regulates development pathways throughout the life cycle of Arabidopsis. Proc Natl Acad Sci USA 108:6680–6685

Dirks-Mulder A, Butôt R, van Schaik P et al (2017) Exploring the evolutionary origin of floral organs of *Erycina pusilla*, an emerging orchid model system. BMC Evol Biol 17:89

Ditta G, Pinyopich A, Robles P, Pélaz S, Yanofsky MF (2004) The *SEP4* gene of *Arabidopsis thaliana* functions in floral organ and meristem identity. Curr Biol 14:1935–1940

Dodson CH (1957) *Oncidium pusillum* and its allies. Amer Orchid Soc Bull 26:170–172

Duttke S, Zoulias N, Kim M (2012) Mutant flower morphologies in the wind orchid, a novel orchid model species. Plant Physiol 158:1542–1547

Félix LP, Guerra M (2000) Cytogenetics and cytotaxonomy of some Brazilian species of Cymbidioid orchids. Genet Mol Biol 23:957–978

Heyduk K, Hwang M, Albert V, Silvera K, Lan T, Farr K, Chang TH, Chan MT, Winter K, Leebens-Mack J

(2019) Altered gene regulatory networks are associated with the transition from C3 to crassulacean acid metabolism in *Erycina* (Oncidiinae: Orchidaceae). Front Plant Sci 9:2000. https://doi.org/10.3389/fpls.2018.02000

Hsu HF, Hsu WH, Lee YI, Mao WT, Yang JY, Li JY, Yang CH (2015) Model for perianth formation in orchids. Nature Plants 1:15046

Huang JZ, Lin CP, Cheng TC, Huang YW, Tsai YJ, Cheng SY, Chen YW, Lee CP, Chung WC, Chang BCH, Chin SW, Lee CY, Chen FC (2016) The genome and transcriptome of *Phalaenopsis* yield insights into floral organ development and flowering regulation. PeerJ 4:e2017. https://doi.org/10.7717/peerj.2017

Gravendeel B, Dirks-Mulder A (2015) Floral development: Lip formation in orchids unravelled. Nature Plants 1:15056. https://doi.org/10.1038/nplants.2015.56

Lapjit C, Tseng MJ (2015) Effects of LEDs (light-emitting diodes) lights on the in vitro growth of *Erycina pusilla*. Horticulture NCHU 40(2):23–38

Lee S, Li C, Liau C et al (2015) Establishment of an *Agrobacterium*-mediated genetic transformation procedure for the experimental model orchid *Erycina pusilla*. Plant Cell Tiss Organ Cult 120:211–220. https://doi.org/10.1007/s11240-014-0596-z

Li CW, Chan MT (2018) Recent protocols on genetic transformation of orchid species. In: Lee YI, Yeung ET (eds) Orchid Propagation: from Laboratories to Greenhouses—Methods and Protocols. Springer Protocols Handbooks. Humana Press, New York, NY. https://link.springer.com/protocol/10.1007/978-1-4939-7771-0_20

Lin WJ (2012) Studies on the gene transformation of *Erycina pusilla*. MS Thesis, Department of Horticulture, National Chung Hsing University, Taichung, Taiwan

Lin YC, Cheng YM, Chang C (2013) Mutation breeding of *Erycina pusilla* (L.) Williams et al. Horticulture NCHU 38(4):81–94

Lin CS, Hsu CT, Liao DC, Chang WJ, Chou ML, Huang YT, Chen JJ, Ko SS, Chan MT, Shih MC (2016) Transcriptome-wide analysis of the MADS-box gene family in the orchid *Erycina pusilla*. Plant Biotechnol J 14:284–298

Liu KT, Chiu YT, Chang C (2011) Intergeneric cross between *Erycina pusilla* and Oncidiinae species and cultivars. Seed & Nursery (Taiwan) 13:31–39

Melters DP, Paliulis LV, Korf IF, Chan SWL (2012) Holocentric chromosomes: convergent evolution, meiotic adaptations, and genomic analysis. Chromosome Res 20:579–593

Mondragón D, Maldonado C, Aguilar-Santelises R (2007) Life history and demography of a twig epiphyte: a case study of *Erycina crista-galli* (Orchidaceae). Selbyana 28(2):137–144. https://www.jstor.org/stable/41760304

Pan IC, Liao DC, Wu FH, Daniell H, Singh ND et al (2012) Complete chloroplast genome sequence of an orchid model plant candidate: *Erycina pusilla* apply in tropical *Oncidium* breeding. PLoS ONE 7(4):e34738. https://doi.org/10.1371/journal.pone.0034738

Sheehan T, Sheehan M (1989) Orchid genera illustrated 127-*Psygmorchis*. Amer Orchid Soc Bull 58:24–25

Theißen G (2001) Development of floral organ identity: Stories from the MADS house. Curr Opin Plant Biol 4:75–85

Tsai WC, Dievart A, Hsu CC, Hsiao YY, Chiou SY, Huang H, Chen HH (2017) Post genomics era for orchid research. Bot Stud 58:61. https://doi.org/10.1186/s40529-017-0213-7

Yeh HY, Lin CS, Chang SB (2015) Cytogenetic and cytometric analyses in artificial intercytotypic hybrids of the emergent orchid model species *Erycina pusilla*. Euphytica 206:533–539. https://doi.org/10.1007/s10681-015-1534-9

Yeh HY, Lin CS, de Jong H, Chang SB (2017) Two reported cytotypes of the emergent orchid model species *Erycina pusilla* are two different species. Euphytica 213:233. https://doi.org/10.1007/s10681-017-2026-x

Phalaenopsis Genome and Transcriptome Exploitation and Its Application for Breeding

4

Kotapati Kasi Viswanath, Jian-Zhi Huang, Shih-Wen Chin, and Fure-Chyi Chen

Abstract

Orchidaceae is one of the largest families of flowering plants with about 30,000 species. Orchid flowers are highly valued because of their quite distinct and pretty features like spectacular floral shape, color, and elaborate adaptations to pollinators. Among the numerous species, the *Phalaenopsis* genus of about 66 species derived hybrids is the popular potted ornamental plant. Nowadays, there are more commercial demands for the various *Phalaenopsis* hybrids with diverse flower colors, fragrance, disease-free, environmentally protected with long-lasting blooming. To produce *Phalaenopsis* hybrids with diverse features, the whole-genome organization analysis plays a key role to deploy suitable strategies. Genome sequencing has been developed as an important discipline in the plant sciences. During the past decade, remarkable advancement has been made in next-generation sequencing technologies, including 454 pyrosequencing, Illumina/ Solexa technology, and sequencing by oligo ligation detection. To accelerate genetic improvement through molecular breeding, wide-ranging analyses of genes and the genome of *Phalaenopsis* have been conducted using advanced technologies. Whole-genome and transcriptomes were sequenced, assembled, and annotated in (wild-type and peloric mutant plants) *P. equestris, P. aphrodite* and hybrids. The chloroplast DNA sequence information of *P. equestris* and *P. aphrodite* were obtained. Various RNA-Seq reads were produced from flowers and temperature treated parts of *Phalaenopsis* orchids (wild-type and peloric mutant plants) and *P. bellina*. From these produced genome sequences, various databases were constructed for the *Phalaenopsis* genome. We expect *Phalaenopsis* genome sequences, microRNAs, expressed sequence tags, simple sequence repeats, linkage maps, and constructed databases will deliver a valued source for basic and applied research, including comparative genomic analysis, developmental studies and molecular breeding on *Phalaenopsis*.

Abbreviations

Myr	Million years
CAM	Crassulacean acid metabolism
NGS	Next-generation sequencing
SOLiD	Sequencing by oligo ligation detection
PGM	Personal genome machine

K. K. Viswanath · J.-Z. Huang · S.-W. Chin · F.-C. Chen (✉)
Department of Plant Industry, National Pingtung University of Science and Technology, Pingtung, Taiwan
e-mail: fchen@mail.npust.edu.tw

F.-C. Chen
General Research Service Center, National Pingtung University of Science and Technology, Pingtung, Taiwan

F.-C. Chen and S.-W. Chin (eds.), *The Orchid Genome*, Compendium of Plant Genomes,
https://doi.org/10.1007/978-3-030-66826-6_4

NCBI	National Center for Biotechnology Information
TCV	Total chromosome volume
CH	Constitutive heterochromatin
ESTs	Expressed sequence tags
BESs	Bacterial artificial chromosome end sequences
TEs	Transposable elements
LTRs	Long terminal repeats
DEGs	Differentially expressed genes
GO	Gene Ontology
KEGG	Kyoto Encyclopedia of Genes and Genomes
FISH	Fluorescent in situ hybridization
sRNA	small RNA

4.1 Introduction

Orchidaceae is the biggest family of angiosperms, with 880 genera and around 30,000–30,315 species. Newly discovered species are reported elsewhere cumulatively. They occupy approximately 30% of monocotyledons and 8% of all vascular plant species (Atwood 1986; Hsiao et al. 2011a, b). Orchids have a unique place in floricultural crops that are commercially grown worldwide. Orchidaceae comprises many distinctive physiological features such as large and complex genomes, zygomorphy, specialized pollination, lacking endosperm, epiphytism, slow growth and long lifespan, and ecological strategies (Leitch et al. 2009; Su et al. 2011). Orchids are enormously rich in species, and the speciation rate is extremely high compared with that of most angiosperms, which suggests that orchids are still successfully evolving (Gill 1989). Since Darwin's time, many biologists showed strong attention and never lost their fascination with the evolution of orchids. Currently, information about genetic sequences is giving insights into the quick evolutionary pattern of orchids. Ramirez et al. 2007 constructed a phylogenetic tree centered on retrieved plastid DNA sequences for 55 orchid genera and the constructed phylogenetic tree was standardized by executing a relaxed-clock model. According to these phylogenetic analyses and plant-pollinator interaction studies, it was found that the utmost current communal ancestor of current orchids existed in the Late Cretaceous (76–84 Myr), and the stem lineages of orchid subfamilies were present formerly in the end of the Cretaceous, ~65 Myr ago (Ramirez et al. 2007). Recently, extraordinary diversification of orchids was resolved for 39 species based on 75 chloroplast genes and found that orchids seem to have arisen approximately 112 Myr ago (Givnish et al. 2015).

Conforming to the horticultural point of view, orchids are placed in the top position in angiosperms with greater commercial value due to their extraordinary floral diversification, including the importance of pharmaceutical and fragrance industries. They have been considered the 'queen of flowers' because of their diversity in shape, size, and color. The juvenile period in orchids is generally between 3 and 13 years and it can vary based on several factors such as environmental conditions and the genetic setup of the genus, species, or hybrids (Ziv and Naor 2006). However cultivation and fertilizer management may shorten the period. The zygomorphic structure of orchid flowers in contrast to most plant groups leads the style of pollinating insects to the flowers from a specific direction (Cubas 2004). Orchid flowers generally contain three sepals in outer whorl, two lateral inner petals in inner whorl, and a specialized labellum. The labellum is usually displayed as a diverse morphology and offers pollinators a landing platform for mimicry features. Sepals and petals are generally called perianths; subsequently, they have regular shape and texture and perform both protective and attractive functions. The gynoecium and androecium, which are inner whorls of floral organs, are fused into a single structure called the gynostemium present in the center (Dressler 1993). The ovary is inferior concerning the rest of the organs, whose maturation is initiated by pollination.

Orchids are found in nearly every habitat in the world except at the glaciers and deserts. Most of these plants are either terrestrial, epiphytic, semiaquatic, or hemiepiphyte. These are extensively dispersed in the tropics and subtropics of Asia, North, Central, and South Americas and temperate regions of Asia and Europe. As revealed by phylogenetic studies, Orchidaceae consists of five subfamilies, including Apostasioideae, Cypripedioideae, Vanilloideae, Orchidoideae, and Epidendroideae (Chase et al. 2003). According to the Royal Horticultural Society in the last 150 years, approximately 110,000 artificial hybrids are recorded and each year 3,000 new orchid hybrids are listed (Hinsley et al. 2017). According to the reports, during 1996–2015 the most marketable orchid trade was from affected proliferated sources, comprising 99.9% of >1.1 billion live orchid plants in commerce and >31 million kg of stems; particularly, Taiwan and Thailand were the leading exporters (UNEP-WCMC 2017). From 2007 to 2012, the typical value of fresh cut orchid trade was US $483 million (De et al. 2014).

The most common type of orchid is perhaps *Phalaenopsis* or commonly named moth orchid. *Phalaina* means 'moth' and *opsis* means 'appearance' in Greek. The *Phalaenopsis* genus is classified into five subgenera principally based on plant size and floral morphology. Approximately 66 species are dispersed all over tropical Asia, and despite their classy appearance, low maintenance, and prolonged flowering longevity, these are widespread ornamental plants (Christenson 2001). The *Phalaenopsis* species such as *P. aphrodite* and *P. equestris* are significant moth orchids with various commercial traits. The *Phalaenopsis* has unique characters, including crassulacean acid metabolism (CAM), epiphytic in native habitats in a continuously moist environment, symbiosis with fungi, immature embryos, and highly evolved flowers with zygomorphic floral structure. These are grown in a warm environment like to stay in the 21–30 °C/16–20 °C range during the day/night. Hybrids of the genus *Phalaenopsis* are the most popular orchids with great commercial worth for the floral industry (Lin et al. 2016). Compared with other tropical countries, the motherland of these orchids is Taiwan and has a fairly successful orchid industry.

4.2 Genome Analysis Techniques and Their Importance in Orchid Biology

Genome is the basis of living organisms and deciphering the genome sequence has become an essential resource in biology. Nowadays, whole-genome sequencing, computer-assisted processes, and various genome databases produce an outstanding genetic information source (Varshney et al. 2009). It aids to illustrate individual genomes, transcriptomes, and variations in the genetic level of populations and offers genetic construction related to each trait. During the past decade, next-generation sequencing (NGS) technologies are becoming the greatest prevalent techniques in genomics. These are covering various fields of research from agriculture to clinical diagnostics. In recent times, remarkable progress has been made in NGS technologies in terms of speed, read duration, and a sharp reduction in per-base cost. The pyrosequencing method, Solexa/Illumina sequencing platform, Ion Torrent, and personal genome machine (PGM) are recently developed advances in NGS (Van Dijk et al. 2014). Several strand-specific RNA-Seq procedures for analyzing transcriptomes and their significant biological functions have been developed (Shalek et al. 2014). The NCBI is accessible for a total of 24,002 genome information, which comprises eukaryotes, prokaryotes, viruses, plastids, and organelles (Valliyodan et al. 2017). The speed of genome sequencing is slow in plants as compared with other living organisms (Valliyodan et al. 2017). Most of the plant genomes are sequenced and published, but these belong to the crop and model plants only; comparatively, only quite a few molecular studies have been dedicated to orchids (Peakall 2007). Because of unique characteristic features, including the specialized pollination strategy, the complexity of genome structures with more species counts in orchids has attracted more attention to study the genomic information. Many researchers are working on the orchid genomics to collect and develop genomic information with the genome and transcriptomic analyses and stored the genetic information in the form of various databases. The present research on the moth orchid genetic information unlocks the doors to discovering orchid diversity and evolution and provides more singificant opportunity to understand orchid biology. The list of various NGS technologies applied for sequencing of *Phalaenopsis* species is presented in Table 4.1. In the current chapter, the present status of large-scale examination of genes and genome of *Phalaenopsis* will be discussed, and their characteristics are summarized.

4.3 Chromosomal Features of *Phalaenopsis*

Over the years, consumers have pursued eternally beautiful and versatile morphological traits in flowers, such as colors, shapes, and fragrance. The *Phalaenopsis* hybrids are among the important potted flowers of commercial value and are produced in large quantities annually. This genus belongs to the subfamily Epidendroideae, tribe Vandeae, and subtribe Aeridinae. In interspecific hybridization and progeny range, wild-type plants are normally used as the parent plants. Because of its high commercial value of *Phalaenopsis* genus, genome organization study is very crucial to molecular geneticists and plant breeders to study and produce various commercially valuable traits with modern technologies. To date, genomes from *Cymbidium, Corallorhiza*, *Phalaenopsis*, *Neottia*, *Oncidium*, *Erycina*, and *Rhizanthella* genera have been sequenced. *Phalaenopsis* species are diploid with 38 chromosomes with various sizes extending from 1.5 to 3.5 μm (Arends 1970; Lin et al. 2001). Commercial *Phalaenopsis* hybrids could be tetraploid or higher ploidy, or even aneuploidy (Kuo et al. 2005; Lee et al. 2020). According to their chromosome sizes, *Phalaenopsis* species can be separated into small and uniform chromosome groups (1–2.5 μm) including *P. aphrodite*, *P. equestris*, *P. stuartiana*, *P. cornu-cervi*, and *P. lueddemanniana*; the small and large chromosomes including *P. venosa*, *P. amboinensis*, and *P. violacea* (Kao et al. 2001). The haploid genome size of *P. equestris* is 1600 Mb (<2 μm long), comparatively higher than rice, grape but similar to sorghum. Lin et al. (2001) assessed the nuclear DNA content of 18 *Phalaenopsis* species and found a six-fold difference between these species and the proposed 2C-values of the *Phalaenopsis* spp. correlated with their chromosome sizes (Lin et al. 2001). A four-fold variance in total chromosome volume (TCV) and a 15-fold difference in the amount of constitutive heterochromatin (CH) were observed in karyotype of nine *Phalaenopsis* species based on somatic metaphase chromosomes (Kao et al. 2001). The genome structure and relationship of the *Phalaenopsis* species have been investigated using genomic in situ hybridization (GISH) (Chih-Chung et al. 2005). Species with large genomes have more repetitive sequences, and small genomes have unique sequences of their own. Recently, the modified drop method has been developed to prepare meiotic chromosomes to research the pachytene spreads from pollinia of two Taiwan native *Phalaenopsis* species, *P. aphrodite* subsp. *formosana* and *P. equestris* (Kuo et al. 2016). These meiotic chromosomes have advantages with less condense and disclose a perfect distribution array of heterochromatic and euchromatic regions. Based on their studies, Kuo et al. (2016) developed the mapping of 45S rDNA and 5S rDNA and also found up to 20-fold higher resolution of chromosomes at the pachytene stages. According to flow cytometry karyomorphology and fluorescence in situ hybridization studies in *Phalaenopsis*, the progression of genome size and different sharing patterns of rDNA have been investigated (Lee et al. 2017). This study reveals that the polymorphism in the number of rDNA loci discloses the complex chromosomal rearrangements. It leads to the genome size increase or decrease and may have led to the generation of the different karyotypes in each *Phalaenopsis* lineage.

Table 4.1 List of various next-generation sequencing (NGS) technologies applied to sequence *Phalaenopsis* genome and transcriptomes

Species	Sequencing platform	Tissue	Purpose of the study	References
P. equestris	Illumina/Solexa	Mature flower buds	Gene expression levels in reproductive organs	Tsai et al. (2006)
P. aphrodite *P. equestris* *P. bellina*	Sanger: EST Roche/454	Protocorm Mature flower buds Mature flower buds without column	Gene discovery and genomic annotation	Hsiao et al. (2011a, b)
P. equestris	Illumina/Solexa	Leaves	Expressed genes involved in many biological processes of orchids	Su et al. (2011)
P. aphrodite	Illumina/Solexa; Roche/454	Leaves, stems, roots, young inflorescences, stalks, flower buds, flowers, germinating seeds	Expressed genes involved in many biological processes of orchids	Su et al. (2011)
P. aphrodite	Illumina/Solexa	Leaves, stalks, flower bud	Roles of small RNAs on the flower regulation under low ambient temperature	An and Feng-Ming (2012), An et al. (2011), Huang et al. (2016)
P. aphrodite	Illumina/Solexa	Mature leaves, roots, flowers, and germinating seeds	Identify species-and tissue-specific miRNAs	Chao et al. (2014)
P. Brother Spring Dancer 'KHM190'	Illumina Hiseq 2000	Petals, sepals or labellum from flower buds of wild type and peloric petal mutant plants	Floral organ development	Huang et al. (2015)
P. equestris	Illumina/Solexa	Leaves and flowers from several individuals of an inbred line	Gene duplication, CAM pathway, MADS-box genes	Cai et al. (2015)
P. equestris	Illumina HiSeq 2000	Leaves, stem, roots and flowers	Origin and mechanism of CAM pathway	Zhang et al. (2016)
P. aphrodite	Illumina/Solexa	Fresh leaves	Analyze an important species for breeding	Tsai et al. (2015)
P. aphrodite	BigDye terminator cycle sequencing (not NGS technology)	Leaves	Comparative analysis of evolutionary rate	Chang et al. (2006)
P. equestris	Roche 454 GS/FLX pyrosequencing	BAC clones containing chloroplast genome	Comparative analysis with *P. aphrodite* and *Oncidium* Gower Ramsey	Cheng-Fong et al. (2012)
P. aphrodite	Ion Proton and Illumina Hiseq 2000	Leaves or flowers	RNA editing of whole-plastid transcript in leaf and flower	Chen et al. (2017)
Phalaenopsis 'Tiny Star'	Illumina MiSeq platform	Leaves	Comparative analysis with the reported orchid chloroplast sequences	Kim et al. (2016)

4.4 *Phalaenopsis* Genome and Transcriptome Sequence Analyses

Angiosperms have various genome sizes. Principally monocots have large genomes. Among these, Orchidaceae perhaps is the most diverse flowering family with the genome sizes extending 168-fold (1C = 0.33–55.4 pg). The remarkable progress in genome sequencing technologies and associated bioinformatics developments have permitted us to obtain whole genomes of sequence information in a comparatively short period. The development of transcriptome sequences of *Phalaenopsis* plays a vital role in orchid biology. The significance of the whole-genome sequencing in plant genomics came to know for the first time after the sequencing of the whole genome of *Arabidopsis thaliana* in 2000 (Arabidopsis Genome Initiative 2000).

4.4.1 *Phalaenopsis* Genome Sequence Analysis

The information concerning orchid genomic sequences is currently limited. The identified expressed sequence tags (ESTs) derived from cDNA sequences deliver sequencing information to find candidate genes for orchid floral development and in flowering time determination (Tsai et al. 2006, 2007). But this information has not yet a wide-ranging explanation of expression levels of all genes in orchids. However, we still have a shortage of basic sequence information, including the organization and structural pattern of the *Phalaenopsis* genome. The whole-genome sequencing information provides the complete depiction of the expressed genes. The studies on young leaves of *Phalaenopsis* have shown that genome size is small for *P. aphrodite* (2.80 pg/2C), dispersed across 19 chromosomes (Lin et al. 2001). Lin et al. (2001) detected increased nuclear DNA content of TSRI cultivars *P. aphrodite*, W01-38, W01-41, and W01-22 in the ratio of 2:3:4 due to an increase in chromosome number, and the identified nuclear DNA contents have a linear relationship with their ploidy levels. The genome size of two native *Phalaenopsis* species such as *P. equestris* and *P. aphrodite* subsp. *formosana* was estimated and revealed a small genome size of 1.6 and 1.4 pg/1C (Chen et al. 2013, 2014). Lee et al. (2004) observed the ploidy level in *P. aphrodite* subsp. *formosana* 'TS97', *P. equestris* 'W9-51' during flower development, and identified lower ploidy levels in young floral buds than old flowers (Lee et al. 2004). Likewise, the outcomes specified that cells with a greater ploidy level have lesser transition rates and a smaller amount possible for promoting endoreduplication.

The first overview of the *P. equestris* genome was analyzed using bacterial artificial chromosome (BAC) end sequences (BESs) (Hsu et al. 2011). By analyzing two genomic BAC libraries with 5,535 BESs, the information about 950 simple sequence repeat (SSR) or microsatellite content, repeat element composition (23.0%), GC content, DNA transposons (10.8%), and protein-encoding regions (11.6%) were observed in *P. equestris*. The first breakthrough was sequencing a miniature breeding parent species *P. equestris* using a whole-genome shotgun and Illumina technology (Cai et al. 2015). Its genome size was estimated to be 1.16 Gb which is smaller than earlier estimates with 62% repetitive DNA, normally transposable elements (TEs). Among the TEs, long terminal repeats (LTRs) were the maximum copious ~46% of the genome. The total number of assembled scaffolds are 236,185, comprising the 6,359 longest scaffolds. After the construction of the protein-coding gene model, the total predicted protein-coding genes and alternatively spliced forms are 29,431 and 9,021, respectively. The most significant finding is that *P. equestris* is the CAM plant, for which the genome was pioneerly sequenced (Cai et al. 2015) after pineapple (Zhang et al. 2014).

Even though the draft genome of *P. equestris* is rather fragmented, scaffolds are inaccurately assembled. This may limit the breeding process through genome-assisted methods and their efficacy for cross-genome comparisons. But recently sequenced *P. aphrodite* genome clarifies the issues observed in the genome of *P. equestris*. Using whole-genome shotgun sequences the obtained data was integrated with an SNP-based high-density genetic linkage map and a high-resolution pachytene karyotype to the 19 chro mosomes of *P. aphrodite* using FISH (Chao et al. 2018). The assembled chromosomes of *P. aphrodite* genome contain a total length of all scaffolds 1025.1 Mb, and a total of 28,902 protein-coding genes. The assembled genome comprised 60.3% of repetitive elements, majorly with Gypsy-like elements (69.9%) followed by Copia-like elements (10.6%). The recombination rate is 5.5 cM/Mb with 74 recombination hot-spots distributed through the 19 chromosomes. The GC content is 34.4 and 44.3% for the whole genome and coding regions.

To study the insight of the geographical distribution and phylogeny in the *Phalaenopsis*, 50 species of four subgenera nuclear DNA content were measured from ovary tissues (pedicel) before pollination (Chen et al. 2013). The 2C content of these 50 species was extremely inconstant within a range of 2.77–17.47 pg. This genomic analysis delivers information to breeders to escape infertility in interspecific and intraspecific hybridization of *Phalaenopsis* species with its genomic constitutions may contain clues to breed for profitable varieties. The draft genome has also been produced from a hybrid *Phalaenopsis* Brother Spring Dancer 'KHM190', an important winter flowering cultivar (Huang et al. 2016). The filtered high-quality 300.5 Gb (90-fold coverage) of genome sequence data of 'KHM190' cultivar assembled with the Velvet assembler. The estimated genome size of 'KHM190' cultivar is 3.45 Gb and the total size of assembled contigs through de novo assembly leads to a sequence of 3.1 Gb, signifying 89.9% of the *Phalaenopsis* orchid genome. The assembled genome has around 59.74% of repetitive elements, comprising LTRs (33.44%), DNA transposons (2.91%) and unclassified repeats (21.99%). The GC content is 30.7%, and the number of gene models are 41,153. Huang et al. (2016) also produced a draft genome for *Phalaenopsis pulcherrima* 'B8802' (a summer-flowering species) through the comparative analysis of obtaining genome data between the two *Phalaenopsis* cultivars and identified 691,532 single-nucleotide polymorphisms. The genome sequence information of both summer flowering species and other low ambient temperature inducible spiking may provide clues for the difference in the flowering behavior and extended for better control.

4.4.2 *Phalaenopsis* Transcriptome Sequence Analysis

Transcriptome sequencing covers an extensive range of applications to detect and explore the whole transcriptome, including both coding mRNA and non-coding RNA. These applications, collectively called RNA-Seq, are becoming more accessible and costs are coming down. Though, it is sequencing-based and suitable for specialty applications such as RNA editing and allele-specific expression.

Classical breeding projects have concentrated on the selection of orchid traits over the past decades. However, a deeper understanding of the genomic structure and the expansion of appropriate molecular methods are compulsory measures to increase the performance of breeding programs. Recently, various transcriptome sequences were generated for orchids including *Phalaenopsis* species using NGS technologies. The informative EST dataset and transcriptome were established by using massively parallel 454 pyrosequencing from 10 samples of *Phalaenopsis* plants, including the tissues of young leaf, flower bud, inflorescence, root, old leaf, cold stressed leaf, *Erwinia chrysanthemi*-infected leaf from *P. equestris*; the tissues of protocorm, cold night temperature-induced spike from *P. aphrodite*; day 5 post-anthesis flower tissue from *P. bellina* (Hsiao et al. 2011a, b). From these transcriptomic sequences of *P. equestris*, *P. aphrodite*, and *P. bellina*, a total of 22,234 different genes were identified, among the 780 unigenes encoding putative transcription factors (TFs). RNA-Seq reads were produced from six diverse floral organs of the *Phalaenopsis* Brother Spring Dancer 'KHM190' wild-type and peloric mutant with a lip-like petal by the Illumina HiSeq 2000 platform (Huang et al. 2015). Subsequently generating transcriptome sequences, a total of 43,552 contigs were acquired after de novo assembly.

Transcriptome sequences have been generated for cDNA library preparation from a comprehensive choice of *P. aphrodite* vegetative and reproductive tissues using Roche 454 and Illumina/Solexa sequencing technologies (Su et al. 2011). A total of three cDNA libraries including (1) vegetative library from mature leaf, stem, and root; (2) reproductive library from young inflorescence, mature stalk, full blossoms, flower buds, and senescing flowers; and (3) seed library from a collection of germinating seeds was constructed. For maximizing assembly efficiency, appropriate software like Newbler, Roche 454 reads, CLC Genomics Workbench (CLC bio) have been used to process sequence data. After assembly, 233,823 unique sequences with 42,590 contigs were obtained. The sequence precision of the collected contigs was confirmed to be as high as 99.9%. Among these 42,590 contigs, 7,263 coding genes were annotated to protein-coding genes and established near full length.

The wide-ranging exploration of transcriptome and expression profiles during leaf explant browning of *Phalaenopsis* hybrid Konggangjinli has been analyzed using Illumina high-throughput technology and de novo assembly (Xu et al. 2015). A total of 79,434,350 clean reads was assembled into 31,708 isogenes and 26,565 annotated unigenes generated. Differentially expressed genes (DEGs) were identified before and after explant browning, and 11 genes were selected by analyzing DEGs through the gene ontology (GO) terms, Kyoto Encyclopedia of Genes and Genomes (KEGG) annotations, and potential Pfam domains.

The evolution of the CAM pathway from the C_3 pathway was recognized by a comparison of the six genes encoding the significant enzymes of the CAM pathway and transcriptomes from 13 orchid species and two existing orchid genomes (Cai et al. 2015; Zhang et al. 2016). Up to 40% of the orchid species adopt CAM metabolism to fix CO_2 (Silvera et al. 2009). The first step in CO_2 fixation is reaction converting CO_2 into HCO_3^-; it is catalyzed by the carbonic anhydrase (CA) and principally fixed by phosphoenolpyruvate carboxylase (PEPC). There are two CA subfamilies such as α-CA and β-CA present in *P. equestris*, but the functional difference between α-CA and β-CA is still not clear. The results showed that the evolution of the CAM mechanism from the C_3 pathway in orchids may be due to the transcriptional regulation of key genes of CAM, genes conserved with the same copy number (specifically for PEPC), fewer whole-genome duplications (WGDs) (there are no correlations between the presence of CAM and gene duplication), and extensive gene loss due to quick evolution (Zhang et al. 2016).

Various genes were analyzed from downloaded transcriptome sequences from *Phalaenopsis* database. The core cell-cycle genes and their expression levels were comprehensively analyzed in *P. aphrodite* using transcriptome sequence (Lin et al. 2014). Eight types of cyclin-dependent kinases (CDKs) (A-, B-, C-, D-, E-, F-, G-, and L-types) were identified with significant conservation, particularly CDKA is evolutionarily conserved. By analyzing expression results, Lin et al. (2014) found that core cell-cycle genes active throughout the G1/S, S, and G2/M stages were especially enhanced in the meristematic tissues that have high production. The genome-wide analysis discovered 30 putative phytocyanins genes and 23 TEOSINTE-BRANCHED/CYCLOIDEA/PCF proteins in *P. equestris* (Xu et al. 2017; Lin et al. 2016). Among these genes, nine phytocyanin genes were highly expressed in flowers, stems, and roots; two TCP proteins (PePCF10 and PeCIN8) were expressed in ovules. These findings suggest that these genes play significant roles in the growth and development of *P. equestris*.

4.4.2.1 miRNA Analysis

The small regulatory RNAs and low ambient temperature-responsive miRNAs are widely studied in *P. aphrodite* subsp. *formosana* (An et al. 2011). Four pools of small RNAs (sRNAs) were generated from *P. aphrodite* subsp. *formosana* under untreated, stalks, low-temperature-treated leaves, and flower buds. Small RNA analysis revealed that a total of 37,533,509 sRNA reads generated 11,129 independent orchid miRNA sequences and most with 21-nt in length, signifying 329 known miRNA families. They also identified 14 putative miRNAs precursors of 10 miRNA families and their target transcripts using the established transcriptomic database. Expression studies and comparative analysis of 11 miRNA families from flower buds, leaves, stalks, and low-temperature-treated leaves revealed that miR156, miR162, miR528, and miR535 were induced under low ambient temperature. It has also been determined that the putative function of the conserved miR156 targets *SQUAMOSA PROMOTER-BINDING-LIKE* (*SPL*), a regulatory gene in floral transition and miR172 may promote an intervening transition from the vegetative to the reproductive phase in *Phalaenopsis*. miR162 may mediate the degradation of *DICER* transcripts, but target genes of miR528 and miR535 were not confirmed because their target mRNAs are short-lived. BLASTX analysis with selected target transcripts discloses a great degree of maintenance of miRNAs in *P. aphrodite* subsp. *formosana* and is comparable to that of *A. thaliana*. Most conserved miRNAs are potentially complicated in the flowering regulation and signify that the sequences and functions are evolutionarily conserved. Gene ontology revealed that most of the target transcripts were vastly associated with plant development and metabolic progressions (An et al. 2011).

The novel miRNAs and their regulatory functions of target genes and sRNAs are further examined in *P. aphrodite* subsp. *formosana* by merging the three transcriptome sequence libraries such as transcriptome, sRNAs, and the degradome (An and Ming 2012). Degradome analysis delivered data of about 15 conserved miRNA families with 29 target genes. After this further deep sequencing of the transcriptome, 22 conserved miRNA and 471 putative novel miRNA precursors were found. Among these 471 novel miRNA precursors, initially, 14 putative novel miRNAs candidates with target transcripts were identified, but from annotations, only seven target transcripts could be found and these novel miRNAs are functionally involved in several developmental phases and stress reactions. Moreover, through degradome analysis, 46 copious sRNA groups were identified. Among them, 25 are up-regulated and 13 down-regulated by low ambient temperature treatment. Target genes of sRNAs are functionally involved in the plant stress response and metabolic reactions.

The miRNA transcriptome was generated using the Illumina Genome Analyzer IIx system and characterized for identifying species- and tissue-specific miRNAs in *P. aphrodite* (GenBank: SRA050114) (Chao et al. 2014). Four small RNA libraries were created using transcriptome sequence data from a collection of mature leaves, roots, flowers, and germinating seeds. A total of 59,387,374 reads, with 1,649,996 being unique reads, were attained from four small RNA libraries. A total of 3,251 unique reads for 181 known plant miRNAs from 88 miRNA families, 23 new miRNAs, 91 precursors, and 240 miRNA targets were recognized through a designed systematic bioinformatics analysis pipeline. Compared with An et al. (2011) analysis, Chao et al. (2014) identified additional 36 known miRNA families related to the tissues and developmental stages. Among these 36 miRNAs, miR3440, miR774, miR4221, miR5654, and miR2950 are highly conserved with *Arabidopsis lyrata*, *A. thaliana*, and *Helianthus annuus*. Moreover, miR2868, miR2905, miR2931, miR5155, miR5532, and miR5538 are conserved with *Oryza sativa*. Tissue-specific differential expression analysis indicated that the miR159 family in *P. aphrodite* displayed all the miRNA families' maximum expression levels across all tissue types. From the RNA-Seq data of *Phalaenopsis* Brother Spring Dancer 'KHM190', conserved miRNAs predicted using Bowtie 2 software (Langmead and Salzberg 2012), and INFERNAL (v 1.1) software (Nawrocki and Eddy 2013) were used to predict miRNA precursor sequences from 92,811,417 sRNA raw reads (18–27 bp) (Huang et al. 2015). It signified 6,976,375 unique sRNA tags. A total of 650 miRNAs belonging to 188 conserved miRNA families and 1,644 miRNA-targeted genes were expected from 96 out of 188 miRNA families. Non-coding RNAs such as rRNAs, tRNAs, snoRNAs, and snRNAs were also identified (Huang et al. 2015).

4.4.2.2 ESTs Analysis

High-scale EST sequencing delivers access to the genome of organisms and it reveals enormous evidence hidden in the genome-scale expression data. Before the development of NGS technologies, Sanger technique was used for the EST sequencing. To observe the somaclonal variations, the genetic difference of *Phalaenopsis* hybrids was studied through the 209 ESTs from wild-type plants and 230 ESTs from peloric flower buds derived from randomly amplified polymorphic cDNAs (cDNA-RAPD) (Chen et al. 2005). The sequenced ESTs showed their involvement in diverse cellular processes together with metabolism, transcription, signal transduction, cell growth and division, protein synthesis, and protein localization. Expression analysis of selected ESTs displayed that cyclophilin, TCP-like genes, and auxin-regulated protein kinase were up-regulated, and a *bZIP* TF was down-regulated in the peloric mutant. The wild-type flower buds have enhanced ESTs than mutant flower buds and are predominantly involved in DNA methylation, chromatin remodeling, and posttranscriptional regulation.

To explore the gene expression in floral organs, a total of 5,593 ESTs with an average 590 bp length were randomly selected and functionally characterized from the cDNA library of mature flower buds of *P. equestris* (Tsai et al. 2006). Using PHRAP program (http://www.phrap.org) 5,593 ESTs were assembled into unigene set of 3,688 sequences (comprising 732 contigs and 2,956 singletons). Expression analysis revealed that the maximum expressing level for transcripts of the *P. equestris* flower buds was homologous to RNA-dependent RNA polymerase of *Cymbidium mosaic virus* (*CymMV*), which could be due to viral infection, followed by heat shock-related protein (HSP) and β-1,3-glucanase. Comparative investigation of the ESTs of *P. equestris* flower buds with assembled ESTs from floral tissue of *A. thaliana* (2707 ESTs), *Oryza sativa* (3934 ESTs), *Acorus americanus* (4548 ESTs), and *Asparagus officinalis* (4267 ESTs) showed that incidences of ESTs allotted to expected functions were very comparable across these species. Floral transcriptomes were similar among the angiosperms from basal monocots to monocots and eudicots. Also, 2,359 ESTs were sequenced and collected into 1,187 unigenes from scented *P. bellina* flower buds without the tissue column (Hsiao et al. 2006). By analyzing these ESTs, glyceraldehyde-3-phosphate, geraniol, and linalool, a set of floral scent biosynthetic pathway related genes were identified and their expression pattern was compared with a scentless species *P. equestris*.

A total of 22,234 different genes were identified from 206,960 ESTs of 10 samples from *P. equestris*, *P. aphrodite*, and *P. bellina*: the tissues of inflorescence, flower bud, root, young leaf, old leaf, cold stressed leaf, *Erwinia chrysanthemi*-infected leaf from *P. equestris*; the tissues of protocorm, cold night temperature-induced spike from *P. aphrodite*; day 5 post-anthesis flower tissue from *P. bellina* (Hsiao et al. 2011a, b). Among these 22,234 different genes, 780 unigenes encoding putative TFs were identified through GO, KEGG, and gene family analysis. The EST expression analysis showed the transcript abundance of cytochrome P450-like protein precursor, followed by triple gene block 3 (a viral gene), and cytochrome P450 monooxygenase. High expression of these genes specifies that these are essential for orchid existence and adaptation to the environment. Numerous genes of *CymMV* were also found, comprising 30 unigenes of RNA-dependent RNA polymerase, 24 of coat protein, 4 of triple gene block 1, and 2 of triple gene block 2. The virus sequences found in several studies could be due to viral infection, which might mask the detection of some scarcely expressed transcripts of plant origin.

4.4.2.3 Protein Coding Genes

Based on the *Phalaenopsis* genome sequence information, studies of the gene families with significant biological functions have been carried out. Chao et al. (2018) annotated 28,753 protein-coding genes in *P. aphrodite*, comprising 1615 TFs and 1185 transporter genes. Su et al. (2011) identified 8,804 differentially expressed genes in *P. aphrodite* through gene cluster analysis, most of the proteins with unknown function. Among those identified proteins, glycine-rich RNA-binding protein, thioredoxin, glutathione transferase, glutaredoxin, peroxidase, Chl a/b-binding protein, ubiquitins, proteases, histone H2B, and ribosomal proteins were highly expressed in all tissues and play a significant role in defense and metabolic processes. Cyclophilin homolog, an immunosuppressant in humans had been reported in a *Phalaenopsis* hybrid with expression associated with a peloric mutant phenotype (Chen et al. 2005). A total of 14,732 unigenes were developed from ESTs of *Phalaenopsis* orchids, and transcript expression levels of floral organs were analyzed in *P. equestris* using oligonucleotide microarray (Hsiao et al. 2013). Transcriptomes of different floral organs were compared and 173, 11, and 285 unigenes were greatly expressed in sepal, petal, and labellum, respectively. After annotation of these unigenes through GO, KEGG pathways and TF family found that unigenes involved in lipid, energy, and terpenoid metabolisms are expressively differentially dispersed between tepal and labellum. Interestingly, sepal-dominant unigenes encoding WRKY and labellum-dominant unigenes encoding MADS-box TFs (Hsiao et al. 2013) were found.

DEGs were identified in the hybrid *Phalaenopsis* Brother Spring Dancer 'KHM190' floral organs (Huang et al. 2016). The widespread gene analysis delivers significant comprehensions into the floral organ development and flowering time regulation. A total of 41,153 genes were predicted in 15,855 families from RNA-Seq reads with an average mRNA length of 1,014 bp and a mean number of 3.83 exons per gene. Through GO analysis, 16,034, 27,294, and 16,360 genes were dispersed to the cellular component, biological process, and molecular function categories. Overall, 1,842 Pfam domains were identified through Pfam protein motifs analysis. To screen DEGs between the wild-type plant and peloric-mutant flower tissues, Huang et al. (2016) applied DEGseq software (Wang et al. 2010), and found a total of 1,838 genes between the peloric sepal (PS) and wild-type sepal (NS) libraries, 758 genes between the peloric petal (PP) and wild-type petal (NP) libraries, and 1,147 genes between the peloric labellum (PL) and wild-type labellum (NL) libraries were significantly differentially expressed. Among the annotated unigenes, 86 related to flowering time regulation were noticed, including photoperiod pathway genes such as GIGANTEA (GI), PHYTOCHROME INTERACTING FACTOR 3 (PIF3), LATE ELONGATED HYPOCOTYL (LHY), PHYTOCHROME A and B (PHYA, PHYB), and CONSTANS (CO); vernalization pathway genes such as VERNALIZATION (VRN), HETEROCHROMATIN1 (LHP1) and FRIGIDA (FRI); autonomous pathway genes including FCA, FPA, FLOWERING LOCUS KH DOMAIN (FLK) and LUMINIDEPENDENS (LD); floral integrator pathway genes related to AP1, AP2, AGAMOUS (AGL), FLOWERING LOCUS T (FT), FRUITFULL (FUL) and LEAFY (LFY); and GA signaling pathway genes related to GIBBERELLIN BIOSYNTHESIS GENES, GIBBERELLIN RECEPTOR (GID), DELLA domain, and GAMYB.

A total of 3,309 TFs representing 6.34% of the 41,153 predicted protein-coding loci were noticed based on the rules of PlantTFDB v3.0 (Jin et al. 2014) (http://planttfdb.cbi.edu.cn/). The most highly represented TF families are the basic helix-loop-helix (279 genes), AP2/EREBP (271 genes), NAC (224 genes), MYB-related (211 genes), and MYB (179 genes) families. A total of 57 TCP and 164 WRKY gene family members are expressed in diverse tissues of

Phalaenopsis. In *P. equstris* genome 29,431 predicted protein-coding genes were grouped into 3,694 gene families comprising at least two orchid genes. A total of 51 recognized functional MADS-box genes and 9 pseudogenes were recognized (Cai et al. 2015).

4.4.2.4 Transposable Elements

TEs identification provides information related to orchid diversity and evolution, in addition to developmental study. Only limited studies on orchid TEs are available, such as reported in the genome of *Phalaenopsis* Brother Spring Dancer 'KHM190' at the protein and DNA levels (Huang et al. 2016). RepeatProteinMask and RepeatMasker tools were applied at the protein and DNA levels for analysis. A total of 1,598,926,178 TEs were identified in LTR retrotransposons (33.44%) mainly with Gypsy-type (24.43%), Copia-type (4.43%), DNA transposons (2.91%), and unclassified repeats (21.99%). The assembled *P. equstris* genome showed that 59% of the genome is engaged by TEs, including 46% by LTRs (Cai et al. 2015).

4.4.2.5 Simple Sequence Repeats

Analysis of simple sequence repeats (SSRs) is the most extensively applied class of molecular markers used in genetic studies, which provides information about evolution and genetic diversity in orchids. Genome analysis of *Phalaenopsis* Brother Spring Dancer 'KHM190' through MIcroSAtelitte (MISA) (http://pgrc.ipk-gatersleben.de/misa/) has identified a total of 532,285 SSRs (Huang et al. 2016). The predicted SSRs were classified into di-, tri-, tetra-, penta-, and hexa-nucleotide motifs based on tandemly arranged copy number. Among these, dimers (79.71%) and trimers (15.75%) were the most abundant SSRs (Huang et al. 2016).

Through the analysis of *P. equestris* BESs, a total of 950 SSRs were identified with 17.2% occupation and an average distance of 4.8 kb between SSRs (Hsu et al. 2011). Among these identified SSRs, dinucleotides were the maximum copious repeat motifs (34.37%) followed by mononucleotides (29.9%), penta-nucleotides (13.5%), and tri-nucleotide repeats (11.0%), respectively. Among the dinucleotide repeats, AT/TA dimer repeats were the maximum repeated SSRs, signifying 253 (26.6%) of all recognized SSRs. Using RNA-Seq reads of *P. aphrodite* subsp. *formosana*, various EST-SSRs loci were identified and EST-SSRs markers were developed to identify various cultivars in *Phalaenopsis* (Tsai et al. 2015). After SSRs mining, a total of 1,439 EST-SSRs were found with most copious of the di-nucleotide repeats (51.49%), followed by tri (35.23%), hexa- (8.41%), tetra- (4.10%), and penta-nucleotide repeats (0.76%).

4.4.3 Organelle Genome of Orchids

Plastid genomes provided very crucial information for metabolism, systematics, and species identification. The organelle genome of 11 orchid species, including *Phalaenopsis, Oncidium, Dendrobium, Cymbidium, Erycina, Neottia,* and *Cypripedium macranthos*, was sequenced and analyzed. Quite limited evidence is accessible on plastid genomes in the Orchidaceae. Most chloroplast genomes display resemblances in structure as well as gene order and content; however, dissimilarities were found in the association of the inverted repeat/small single-copy junction and *ndh* genes. The chloroplast genomes (cpDNA) of the *P. aphrodite* (NCBI accession number AY916449) and *P. equestris* (accession No. JF719062) have been entirely sequenced and phylogenies of *Phalaenopsis* species also have been conducted (Chang et al. 2006; Cheng-Fong et al. 2012).

The cpDNA of *P. aphrodite* was comparatively analyzed with grasses to study the initial variance of angiosperms, evolutionary rate, and phylogenetic consequences. The cpDNA of *P. aphrodite* consisting of 148,964 bp with a circular nature has a relatively short single-copy region (11,543 bp) due to the infrequent loss and dispersed deletion of definite *ndh* subunits (Chang et al. 2006). In *P. aphrodite*, the miscellanies of the *ndhB*, *ndhC*, *ndhD*, *ndhE*, *ndhG*, *ndhI*, *ndhJ*, and *ndhK* sequences were found which are non-functional. The *ndhA*, *ndhF*, and *ndhH* genes were absent and in-frame sequences of these genes were confirmed by using total DNA PCR assay. Based on these PCR assays, Chang et al. (2006) proposed that the ancestral functional *ndh* copies of the chloroplast genome may have been transferred to the nuclear genome. Recently, Lin et al. (2017) comparatively analyzed the connected loss of NDH complex-related genes within nuclear and chloroplast genomes in *Apostasia odorata*, *P. equestris*, *D. catenatum*, and *D. officinale* orchids. They identified all missed cp-*ndh* genes that are not present in any of these four nuclear genomes. Further, they recognized that the NDH complex can be lost in photoautotrophic plant species. According to cpDNA sequence analysis, a relationship of nucleotide substitutions between *P. aphrodite* and the grasses define that only the plastid expression genes have a solid positive correlation between non-synonymous (Ka) and synonymous (Ks) substitutions per site, given that confirmation for a generation time effect, mostly through these genes. These studies produced strong maintenance for the *Amborella/Amborella* Nymphaeales-basal hypothesis and for the orchid and grasses composed of being a monophyletic group nested within the remaining angiosperms.

The cpDNA of *P. aphrodite* encodes 110 diverse known genes, comprising 74 identified protein-coding genes, and two retained reading frames of unknown function, 4 rRNA genes, and 30 tRNA genes. Also, it listed 24 open reading frames (ORFs) with a threshold of 225 bp. For 74 protein-coding transcripts of the chloroplast genome, RNA editing sites were thoroughly examined and compared with those of other seed plants (Zeng et al. 2007). A total of 44 editing sites have been identified with changes in amino acids being carried out among these 42 editing sites. One of the remaining two editing sites exists in the *ndh*B pseudogene's transcript, and another in the 5' UTR of *psbH* transcripts. Besides, 21 editing sites are unique to *P. aphrodite* as opposed to other seed plants. Recently, the RNA editing status of whole-plastid transcripts in *P. aphrodite* leaf and floral tissues has been recognized using RNA-Seq technology (Chen et al. 2017). A total of 137 edits were identified, including 110 and 106 edits present in leaf and floral tissues. Among these 137 edits, 79 edits were involved in protein-coding transcripts with changes in amino acids, and the 58 modifications of nucleotide caused the non-synonymous substitution. These influence the formation and stability of the secondary structure and perhaps play a significant role in controlling gene expression. The maternal inheritance array of cpDNA was identified in interspecific hybrids of *Phalaenopsis* and intergeneric hybrids of *Phalaenopsis* and *Doritis* (now reclassified as *Phalaenopsis*) by using restriction fragment length polymorphisms (RFLPs) visualized with a *rbc*L probe (Chang et al. 2000). RFLP analyses showed that *P. amabilis*, *P. aphrodite*, and *P. stuartiana*, which belong to the taxonomic *Phalaenopsis* section, have the same size 2.0-kb fragment. Both *P. mannii* and *P. amboinensis* have a 2.3-kb fragment, whereas *Doritis* (*Phalaenopsis*) *pulcherrima* has a 3.5-kb fragment. Fifteen extremely variable cpDNA regions of *P. equestris* cultivars with polymorphic information contents ≥ 8.0 were recognized through the assessment of InDel regions by cpDNA analysis (Liu et al. 2016). These outcomes proposed that the cpDNA can be used as a marker to identify relationship and phylogeny taxonomic studies. Gene-dependent DNA markers were established for the *Phalaenopsis* KS Little Gem, a new *Phalaenopsis* variety dependent on a genome-wide comparison of chloroplast sequences. Sequence assessments of cpDNA from *P.* KS Little Gem with *P. aphrodite* and *P. equestris* identified six variable gene regions of the chloroplast genome in which an intron of *rpl16* was the most polymorphic among similar orchid species (Kwon et al. 2017).

The cpDNA of *P. equestris* with 148,959 bp was split by a pair of inverted repeats (25,846 bp) into large single-copy (85,967 bp) and small single-copy (11,300 bp) regions. The genome encodes 109 genes, comprising 75 protein-coding genes, 4 rRNA genes, and 30 tRNA genes. A + T content occupied 63.35%, 21 ORFs with a threshold of 225 bp, and a total of 43 RNA editing sites in protein-coding transcripts were identified. Comparative analysis of *P. equestris* cpDNA with cpDNA of *P. aphrodite* and *Oncidium* Gower Ramsey disclosed that the coding genes, IR junction, and RNA editing sites between the two moth orchids are comparatively more conserved than those between *Phalaenopsis* and *Oncidium* spp. (Cheng-Fong et al. 2012). The *ndhA*, *ndhE*, *ndhF*, and *ndhH* genes were absent, and the remnants of *ndhB*, *C*, *D*, *G*, *I*, *J*, and *K* genes also had varying degrees of Indels and frameshifted.

The complete chloroplast genome of *Phalaenopsis* 'Tiny Star' was sequenced using the Illumina MiSeq platform (GenBank accession number, KJ944326) (Kim et al. 2016). The assembled cpDNA with 148,918 bp in length, 36.7% complete GC content, 70 protein-coding genes, 30 tRNA genes, and 4 rRNA genes present in the circular form. Comparative sequence analysis of other chloroplast genomes of orchids discovered six distinct variants of InDel in cpDNA called the 'Tiny Star'.

4.5 Databases

After sequencing the *Phalaenopsis* genome and transcriptomes, the sequence information is combined and deposited in databases. Sometimes these databases are updated with newly produced sequences. Constructed databases are beneficial to study the orchid-gene annotations. At present, very inadequate information on the *Phalaenopsis* genome, RNA, and small RNA sequence is accessible. OrchidBase, a web-based EST database (http://lab.fhes.tn.edu.tw/est.) constructed with 37,979,342 sequence reads, consists of 11 *Phalaenopsis* orchid cDNA libraries (protocorm, mature flower bud, mature flower bud without column, seed after imbibition, inflorescence, developing seed, vegetative tissue, cold-treated plantlet, pathogen-treated plantlet, leaf, and mixed tissues) comprising 41,310 ESTs (Fu et al. 2011). These sequences result in 84,617 non-redundant transcribed sequences, including 8,501 contigs and 76,116 singletons. To increase the genetic information of orchids, OrchidBase is updated to OrchidBase 2.0 (http://orchidbase.itps.ncku.edu.tw/) with 1,562,071 newly added Orchidaceae unigenes broadly from mature flower buds of 10 orchid species including *P. equestris* across five subfamilies of Orchidaceae (Tsai et al. 2013). Orchidstra (http://orchidstra.abrc.sinica.edu.tw), a database with transcriptomes of five orchid species (*P. aphrodite*, *P. equestris*, *P. bellina*, *Erycina pusilla*, *Dendrobium nobile*) and one commercial hybrid (*Oncidium* Gower Ramsey) for orchid functional genomics studies is also available (Su et al. 2013). Orchidstra comprises more than 380,000 unigenes, of which >110,000 are protein-coding genes. The descriptions and functional annotation of protein-coding genes were analyzed through the annotation pipeline using GO, KEGG, and Pfam. One of the significant features of Orchidstra is it comprises miRNAs with extending information of miRNA annotation, precursors, and putative target genes of *P. aphrodite*. Later Orchidstra database was updated to Orchidstra 2.0 (http://orchidstra2.abrc.sinica.edu.tw/) to accommodate the growing amount of orchid transcriptome data (Chao et al. 2017). A total of 510,947 protein-coding gene annotations and 161,826 non-coding transcripts of 18 orchid species belonging to 12 genera in five subfamilies of Orchidaceae was stored in the Orchidstra 2.0.

Recently, we built a PhalDB, a broad database (Lee et al. 2018), covering with information of the genome, transcriptome, and miRNA of the various tissues of *Phalaenopsis* species under cool and warm temperature treatments at different duration. Three databases were constructed with the RNA-Seq raw reads, the sets of assembled unigenes and predicted coding sequences. The annotation results of the aligned unigenes to illustrate the transcriptomes for 11 diverse *P. equestris* tissues signifying the leaf, stem, root, flower buds, sepal, petal, lip, column, and three developmental stages of seeds (Niu et al. 2016).

4.6 Conclusions

In summary, whole-genome sequencing is at the cutting edge of life sciences in the new era. Various orchid genomes have been sequenced and analyzed by applying advanced NGS technologies and genome annotation pipelines. These genome sequences of *Phalaenopsis* species and hybrids should deliver valued information for understanding the genetic systems of the *Phalaenopsis* and other orchid species. The identified *Phalaenopsis* species have the same chromosome number ($2n = 2x = 38$) with various sizes ranging from 1.5 to 3.5 μm. The DNA content of various *Phalaenopsis* species is also identified. *P. equestris* genome size was estimated to be 1.16 Gb with 62% repetitive DNA and a total of 29,431 protein-coding genes. The chromosome assembly of *P. aphrodite* genome contains the total length of 1025.1 Mb, and a total of 28,902 protein-coding genes with 60.3% of repetitive elements. The genome size of *Phalaenopsis* Brother Spring Dancer 'KHM190' of assembled contigs is 3.1 Gb, signifying 89.9% of the *Phalaenopsis* orchid genome. The assembled genome contains around 59.74% of repetitive elements, comprising LTRs (33.44%) and DNA transposons (2.91%). Diverse ESTs were identified and analyzed in floral organs, root, young leaf, old leaf, cold stressed leaf, protocorm, cool night temperature-induced spike to study the genetic variability during growth and development. Various miRNAs including miR156, miR162, miR528, miR535, miR2868, miR2905, miR2931, miR5155, miR5532, and miR5538 and their regulatory roles of target genes and sRNAs are examined in *Phalaenopsis*. The total size of 1,598,926,178 TEs and 532,285 SSRs were identified in *Phalaenopsis* Brother Spring Dancer 'KHM190'. All assembled databases contain information about protein-coding genes, physiological pathways of orchid growth and development, miRNAs, ESTs, SNPs, TEs, SSRs. In the future, the transcripts or non-transcripts of orchids offer the opportunity to study their functions using the novel gene editing technology or transgenics that can be foreseeable.

Acknowledgements The works by authors described in this chapter were supported by grants from the Ministry of Science and Technology (MOST 106-2321-B-020-002) and Agriculture and Food Agency, Council of Agriculture, Taiwan (107AS-7.6.3-FD-Z2, 107AS-7.5.3-FD-Z1 and 108AS-7.5.2-FD-Z1) to FCC.

References

An FM, Ming TC (2012) Transcriptome-wide characterization of miRNA-directed and non-miRNA-directed endonucleolytic cleavage using Degradome analysis under low ambient temperature in *Phalaenopsis aphrodite* subsp. *formosana*. Plant Cell Physiol 10:1737–1750

An FM, Shuan RH, Ming TC (2011) Sequencing-based approaches reveal low ambient temperature-responsive and tissue-specific microRNAs in *Phalaenopsis* orchid. PLoS ONE 6:e18937

Arabidopsis Genome Initiative (2000) Analysis of the genome sequence of the flowering plant *Arabidopsis thaliana*. Nature 408:796–815

Arends JC (1970) Cytological observations on genome homology in eight interspecific hybrids of *Phalaenopsis*. Genetica 41:88–100

Atwood JT (1986) The size of Orchidaceae and the systematic distribution of epiphytic orchids. Selbyana 9:171–186

Cai J, Xin L, Kevin V, Sebastian P, Wen CT et al (2015) The genome sequence of the orchid *Phalaenopsis equestris*. Nat Genet 47:65–72

Chang CC, Lin HC, Pin LI, The YC, Chen HH et al (2006) The chloroplast genome of *Phalaenopsis aphrodite* (Orchidaceae): comparative analysis of evolutionary rate with that of grasses and its phylogenetic implications. Mol Biol Evol 23:279–291

Chang SB, Wen HC, Chen HH, Fu YM, Yih SL (2000) RFLP and inheritance patterns of chloroplast DNA in intergeneric hybrids of *Phalaenopsis* and *Doritis*. Bot Bull Academia Sinica 41:219–223

Chao YT, Chun LS, Wen HJ, Wan CC, Yao CAC et al (2014) Identification and characterization of the microRNA transcriptome of a moth orchid *Phalaenopsis aphrodite*. Plant Mol Biol 84:529–548

Chao YT, Shao HY, Jen HY, Wan CC, Ming CS (2017) Orchidstra 2.0 a transcriptomics resource for the orchid family. Plant Cell Physiol 58:1–13

Chao YT, Wan CC, Chun YC, Hsiu YH, Chih HY et al (2018) Chromosome-level assembly, genetic and physical mapping of *Phalaenopsis aphrodite* genome provides new insights into species adaptation and resources for orchid breeding. Plant Biotechnol J 16:2027–2041

Chase MW, Cameron KM, Barrett RL, Freudenstein JV (2003) DNA data and Orchidaceae systematics: a new phylogenetic classification. In: Dixon KW, Kell SP, Barrett RL, Cribb PJ (eds) Orchid conservation.

Natural History Publications, Kota Kinabalu, Sabah, Malaysia, pp 69–89

Chen TC, Yu CL, Xuewen W, Chi HW, Chih HH et al (2017) Whole plastid transcriptomes reveal abundant RNA editing sites and differential editing status in *Phalaenopsis aphrodite* subsp. *formosana*. Bot stud 58:38

Chen WH, Kao YL, Tang CY, Tsai CC, Lin TY (2013) Estimating nuclear DNA content within 50 species of the genus *Phalaenopsis* Blume (Orchidaceae). Sci Hortic 161:70–75

Chen WH, Kao YK, Tang CY (2014) Variation of the genome size among *Phalaenospsis* species using DAPI. J Taiwan Soc Hortic Sci 60:115–123

Chen YH, Tsai YJ, Hung JZ, Chen FC (2005) Transcription analysis of peloric mutants of *Phalaenopsis* orchids derived from tissue culture. Cell Res 15:639–657

Cheng FJ, Tien CC, Jhong YL, Ting CC, Wen LW, Ching CC (2012) The comparative chloroplast genomic analysis of photosynthetic orchids and developing DNA markers to distinguish Phalaenopsis orchids. Plant Sci 190:62–73

Chih-Chung L, Yao HC, Wen HC, Chi CC, Yen YK (2005) Genome organization and relationships of *Phalaenopsis* orchids inferred from genomic in situ hybridization. Bot Bull Academia Sinica 46:339–345

Christenson EA (2001) Phalaenopsis—a monograph. Timber Press, Portland, Oregon, p 330p

Cubas P (2004) Floral zygomorphy, the recurring evolution of a successful trait. BioEssays 26:1175–1184

De LC, Pathak P, Rao AN (2014) Commercial Orchids. Walter de Gruyter, Warsaw and Berlin, p 301p

Dressler RL (1993) Phylogeny and Classification of the Orchid Family. Cambridge University Press, Cambridge, p 321

Fu CH, Yun WC, Yu YH, Zhao JP, Zhong JL et al (2011) OrchidBase: a collection of sequences of the transcriptome derived from orchids. Plant Cell Physiol 52:238–243

Gill DE (1989) Fruiting failure, pollinator inefficiency, and speciation in orchids. In: Otte D, Ender JA (eds) Speciation and Its Consequences. Sunderland, MA, pp p458–481

Givnish TJ, Daniel S, Mercedes A, Stephanie PL, Steven JH et al (2015) Orchid phylogenomics and multiple drivers of their extraordinary diversification. Proc Royal Soc B 282:20151553

Hinsley A, Hugo JB, Michael FF, Stephan WG, Lauren MG et al (2017) A review of the trade in orchids and its implications for conservation. Bot J Linnean Soc 186:435–455

Hsiao YY, Tsai WC, Chang SK, Tian HH, Hei CW et al (2006) Comparison of transcripts in *Phalaenopsis bellina* and *Phalaenopsis equestris* (Orchidaceae) flowers to deduce monoterpene biosynthesis pathway. BMC Plant Biol 6:1–14

Hsiao YY, Pan ZJ, Chia CH, Ya PY, Yi CH et al (2011a) Research on orchid biology and biotechnology. Plant Cell Physiol 52:1467–1486

Hsiao YY, Yun WC, Shi CH, Zhao JP, Chih HF et al (2011b) Gene discovery using next-generation pyrosequencing to develop ESTs for *Phalaenopsis* orchids. BMC Genom 12:360

Hsiao YY, Tian HH, Chih HF, Shi CH, Yi JC et al (2013) Transcriptomic analysis of floral organs from *Phalaenopsis* orchid by using oligonucleotide microarray. Gene 518:91–100

Hsu CC, Yu LC, Tien CC, Yu LL, Yi TK et al (2011) An overview of the *Phalaenopsis* orchid genome through BAC end sequence analysis. BMC Plant Biol 11:1–13

Huang JZ, Chih PL, Ting CC, Bill CHC, Shu-YC et al (2015) A de novo floral transcriptome reveals clues into *Phalaenopsis* orchid flower development. PloS One 10:e0123474

Huang JZ, Chih PL, Ting CC, Ya WH, Yi JT et al (2016) The genome and transcriptome of *Phalaenopsis* yield insights into floral organ development and flowering regulation. PeerJ 4:e2017

Jin J, Zhang H, Kong L, Ge G, Jingchu L et al (2014) PlantTFDB 3.0: a portal for the functional and evolutionary study of plant transcription factors. Nucleic Acids Res 42:D1182–1187

Kim GB, Youngeun K, Hee JY, Ki BL, Jae HS et al (2016) The complete chloroplast genome of *Phalaenopsis* "Tiny Star". Mitochondrial DNA Part A 27:1300–1302

Kuo YT, Hui LH, Chih HY, Song BC (2016) Application of a modified drop method for high-resolution pachytene chromosome spreads in two *Phalaenopsis* species. Mol Cytogenet 9:44

Kwon YE, Hee JY, Seunghoon B, Goon BK, Ki BL et al (2017) Development of gene-based identification markers for *Phalaenopsis* 'KS Little Gem' based on comparative genome analysis. Hortic Environ Biotechnol 58:162–169

Kao YY, Chang SB, Tsai YL, Chang HH, Yao HC et al (2001) Differential accumulation of heterochromatin as a cause for karyotype variation in *Phalaenopsis* orchids. Ann Bot 87:387–395

Kuo HL, Ke CI, Chen YS, Chin SW, Chen FC (2005) Study on root tip squash and staining techniques of *Phalaenopsis* orchids. J Chin Soc Hortic Sci 51:339–345 (in Chinese)

Langmead B, Salzberg SL (2012) Fast gapped-read alignment with Bowtie 2. Nat Methods 9:357–359

Lee CY, Viswanath KK, Huang JZ, Lee CP, Lin CP, Cheng TC, Chang BC, Chin SW, Chen FC (2018) PhalDB:A comprehensive database for molecular mining of the *Phalaenopsis* genome, transcriptome and miRNome. Genet Mol Res 17(4):GMR18031

Lee HC, Chiou DW, Wen HC, Albert H, Yao HC et al (2004) Dynamics of cell growth and endoreduplication during orchid flower development. Plant Sci 166:659–667

Lee YI, Tseng YF, Lee YC, Chung MC (2020) Chromosome constitution and nuclear DNA content of *Phalaenopsis* hybrids. Sci Hortic 262:109089

Lee YI, Mei CC, Hao CK, Chun NW, Yi CL et al (2017) The evolution of genome size and distinct distribution patterns of rDNA in *Phalaenopsis* (Orchidaceae). Bot J Linnean Soc 185:65–80

Leitch JI, Kahandawala I, Suda J, Hanson L, Ingrouille MJ et al (2009) Genome size diversity in orchids: consequences and evolution. Ann Bot 104:469–481

Lin CS, Jeremy JWC, Chi CC, Hsiao CW, Chen JY et al (2017) Concomitant loss of NDH complex-related genes within chloroplast and nuclear genomes in some orchids. Plant J 90(5):994–1006

Lin HY, Jhun CC, Miao JW, Yi CL, Huang HL et al (2014) Genome-wide annotation, expression profiling, and protein interaction studies of the core cell-cycle genes in *Phalaenopsis aphrodite*. Plant Mol Biol 84:203–226

Lin S, Lee HC, Wen HC, Chi CC, Yen YK et al (2001) Nuclear DNA contents of *Phalaenopsis* species and *Doritis pulcherrima*. J Amer Soc Hortic Sci 126:195–199

Lin YF, You YC, Yu YH, Ching YS, Jui LH et al (2016) Genome-wide identification and characterization of TCP genes involved in ovule development of *Phalaenopsis equestris*. J Exp Bot 67:5051–5066

Liu YC, Bo YL, Jhong YL, Wen LW, Ching CC et al (2016) Evaluation of chloroplast DNA markers for intraspecific identification of *Phalaenopsis equestris* cultivars. Sci Hortic 203:86–94

Nawrocki EP, Eddy SR (2013) Infernal 1.1: 100-fold faster RNA homology searches. Bioinformatics 29:2933–2935

Niu SC, Qing X, Guo QZ, Yong QZ, Wen CT et al (2016) De novo transcriptome assembly databases for the butterfly orchid *Phalaenopsis equestris*. Sci Data 3:160083

Peakall R (2007) Speciation in the Orchidaceae: confronting the challenges. Mol Ecol 16:2834–2837

Ramirez SR, Gravendeel B, Rodrigo BS, Charles RM, Naomi EP et al (2007) Dating the origin of the Orchidaceae from a fossil orchid with its pollinator. Nature 448:1042–1045

Shalek AK, Rahul S, Joe S, John JT, Dave G et al (2014) Single-cell RNA-seq reveals dynamic paracrine control of cellular variation. Nature 509:363–369

Silvera K, Santiago LS, John CC, Klaus W (2009) Crassulacean acid metabolism and epiphytism linked to adaptive radiations in the Orchidaceae. Plant Physiol 149:1838–1847

Su CL, Ya TC, Yao CAC, Wan CC, Chun YC et al (2011) De novo assembly of expressed transcripts and global analysis of the *Phalaenopsis aphrodite* transcriptome. Plant Cell Physiol 52:1501–1514

Su CL, Ya TC, Shao HY, Chun YC, Wan CC et al (2013) Orchidstra: an integrated orchid functional genomics database. Plant Cell Physiol 54:1–11

Tsai CC, Huei CS, Hao VW, Yu SL, Chia HC et al (2015) RNA-seq SSRs of moth orchid and screening for molecular markers across genus *Phalaenopsis* (Orchidaceae). PLoS One 10(11):e0141761

Tsai WC, Hsiao YY, Lee SH, Tung CW, Wang DP et al (2006) Expression analysis of the ESTs derived from the flower buds of *Phalaenopsis equestris*. Plant Sci 170:426–432

Tsai WC, Hsiao YY, Pan ZJ, Chen HH (2007) Analysis of expression of *Phalaenopsis* floral ESTs. In: Chen HH, Chen WH (eds) Orchid Biotechnology. World Sci Publishing, Singapore, pp 145–161

Tsai WC, Chih HF, Yu YH, Yueh MH, Li JC et al (2013) OrchidBase 2.0: comprehensive collection of Orchidaceae floral transcriptomes. Plant Cell Physiol 54:1–7

UNEP-WCMC (2017) CITES trade statistics derived from the CITES Trade Database. UNEP World Conservation Monitoring Centre, Cambridge. https://trade.cites.org

Varshney RK, Nayak SN, Gregory DM, Scott AJ (2009) Next-generation sequencing technologies and their implications for crop genetics and breeding. Trends Biotechnol 27:522–530

Valliyodan B, Suk HL, Henry TN (2017) Sequencing, assembly, and annotation of the soybean genome. In: Nguyen H, Bhattacharyya M (eds) The Soybean Genome. Compendium of Plant Genomes. Springer, Cham, pp 73–82

Van Dijk EL, Hélène A, Yan J, Claude T (2014) Ten years of next-generation sequencing technology. Trends Genet 30:418–426

Wang L, Feng Z, Wang X, Xiaowo W, Zhang X (2010) DEGseq: an R package for identifying differentially expressed genes from RNA-seq data. Bioinformatics 26:136–138

Xu C, Biyu Z, Junmei H, Wen H, Yumei L et al (2015) Genome-wide transcriptome and expression profile analysis of *Phalaenopsis* during explant browning. PLoS One 10:e0123356

Xu L, Wang XJ, Wang T, Li LB (2017) Genome-wide identification, classification, and expression analysis of the phytocyanin gene family in *Phalaenopsis equestris*. Biol Plant 61:445–452

Zhang J, Liu J, Ming R (2014) Genomic analyses of the CAM plant pineapple. J Exp Bot 65:3395–404

Zhang L, Chen F, Zhang GQ, Zhang YQ, Shance N et al (2016) Origin and mechanism of crassulacean acid metabolism in orchids as implied by comparative transcriptomics and genomics of the carbon fixation pathway. Plant J 86:175–185

Zeng WH, Liao SC, Chang CC (2007) Identification of RNA editing sites in chloroplast transcripts of *Phalaenopsis aphrodite* and comparative analysis with those of other seed plants. Plant Cell Physiol 48:362–368

Ziv M, Naor V (2006) Flowering of geophytes in vitro. Propag Ornament Plants 6:3–16

Chromosome Analysis of *Phalaenopsis* Yellow Cultivars

5

Yung-I Lee and Mei-Chu Chung

Abstract

In *Phalaenopsis*, interspecific hybridization has been broadly used for breeding cultivars with new traits of flower colors. Yellow pigments in *Phalaenopsis* cultivars naturally are derived from the diploid species in subgenus *Polychilos*, including *P. amboinensis*, *P. fasciata, P. lueddemanniana* and *P. venosa*, etc. Commercial cultivars of yellow *Phalaenopsis* were made by crossing the diploid species in subgenus *Polychilos* with tetraploid standard-type cultivars, and this often resulted in the formation of triploid progenies. Here we review the breeding history of yellow *Phalaenopsis* cultivars from a cytogenetic perspective. Chromosome analyses indicate polyploidization playing an important role in these breeding programs and reveal different chromosome introgression of section *Polychilos* species in these yellow cultivars. The cytogenetic data of these cultivars could provide basic knowledge for yellow *Phalaenopsis* breeding.

5.1 Introduction

Commercial production of *Phalaenopsis* has increased greatly worldwide as a potted plant during the past few decades because the blooming plants have the long-lasting shelf-life and they are available for sale all year round (Griesbach 2002; USDA 2019). In the ornamental plant breeding, breeders always endeavor to produce more new cultivars with novel traits. Interspecific and intergeneric hybridization are absolutely useful strategies to extend the genomic diversity for the novel traits, such as flower color, flower substance, flower type and vase life in ornamental plants (Van Tuyl and Lim 2003). In *Phalaenopsis* breeding history, the polyploidization has been considered as an important step in producing the standard-type white cultivars (Griesbach 1985). In order to increase the color and patterns of *Phalaenopsis*, novel cultivars with various colors and patterns, such as red, spots, yellow and orange flowers, species from different sections or subgenera were used in the interspecific hybridization (Griesbach1985;Tang and Chen 2007; Chuang et al. 2008). These interspecific hybrids often have low fertility because of their chromosome compositions, e.g. triploidy and aneuploidy (Aoyama 2010), and the

Y.-I. Lee
Department of Biology, National Museum of Natural Science, Taichung 40453, Taiwan, ROC

Department of Life Sciences, National Chung Hsing University, Taichung 40227, Taiwan, ROC

M.-C. Chung (✉)
Institute of Plant and Microbial Biology, Academia Sinica, Taipei 11529, Taiwan, ROC
e-mail: bomchung@gate.sinica.edu.tw

F.-C. Chen and S.-W. Chin (eds.), *The Orchid Genome*, Compendium of Plant Genomes,
https://doi.org/10.1007/978-3-030-66826-6_5

low genomic affinity of parents resulted in abnormal pairing behaviors during meiosis (Arends 1970; Bolanos-Villegas et al. 2008). Various numbers and sizes of chromosomes in the novelty-type cultivars of *Phalaenopsis* have been reported by Aoyama (1993, 2010) and Lee et al. (2020), which suggested the occurrence of polyploidization in these cultivars. Here we review the chromosomal composition of several important cultivars in breeding of yellow *Phalaenopsis*.

5.2 A Cytogenetic Perspective on Breeding Yellow *Phalaenopsis* Cultivars

The breeding and selection of tetraploid white *Phalaenopsis* hybrids, for example *P.* Doris, was the important step in developing standard-type cultivars (Vaughn and Vaughn 1973). Afterward, the tetraploid standard-type cultivars with pink, stripe and semi-alba colors were bred one after another. During 1970s, a number of novelty-type hybrids were made by crossing diploid species in subgenus *Polychilos*, including *P. amboinensis*, *P. fasciata*, *P. lueddemanniana* and *P. reichenbachiana* with standard-type cultivars (Freed 1980a, b, 1981a, b). These species have star-shaped and waxy flowers that contributed various color pigments, such as yellow, spotted and purple to their progenies. Besides, species in subgenus *Polychilos* usually possess a larger chromosome and a bimodal karyotype (Lee et al. 2017). The crosses between tetraploid standard-type cultivars and diploid species usually result in the formation of triploid plants that are generally sterile. We examine a few important cases to demonstrate the role of polyploidization in breeding history of yellow *Phalaenopsis*.

First, *P.* Golden Sands was produced by crossing *P.* Fenton Davis Avant (a tetraploid standard white hybrid) with a diploid species with maroon spots and yellow background color, *P. lueddemanniana* (later it is recognized as *P. fasciata*). The most famous cultivar, *P.* Golden Sands 'Canary' (Fig. 5.1a) received a First Class Certificate from the American Orchid Society. However, *P.* Golden Sands 'Canary' is a triploid plant ($2n = 3x = 57$) with low fertility. After treated with colchicine, a hexaploid ($2n = 6x = 114$) *P.* Golden Sands 'Canary' (Fig. 5.1b) was produced (Griesbach 1985). Subsequently, breeders in the USA used the hexaploid *P.* Golden Sands 'Canary' to cross with diploid species and produced a number of yellow *Phalaenopsis* hybrids, such as *P.* Golden Amboin (*P.* Golden Sands × *P. amboinensis*), *P.* Liu Tuen-Shen (*P.* Golden Sands × *P. gigantea*), *P.* Goldiana (*P.* Golden Sands × *P. lueddemanniana*), and *P.* Golden Bells (*P.* Golden Sands *P. venosa*). These progenies are of great importance because they are tetraploid (or near tetraploid) yellow cultivars with strong color and good fertility, and they have been substantially used for breeding not only yellow but also spotted, orange and red *Phalaenopsis*. A famous green color with red lip cultivar, *P.* Fortune Saltzman 'Maple Bridge' (Fig. 5.1g), a hypotetraploid plant ($2n = 4x - 6 = 70$) (Fig. 5.2b) was selected from the cross between *P.* Liu Tuen-Shen and *P.* Barbara Freed Saltzman.

Another well-known yellow cultivar is *P.* Golden Emperor 'Sweet', which possesses clear yellow flowers without obvious spots (Fig. 5.1e) and is a triploid ($2n = 3x = 57$, Fig. 5.2a). It was selected in Taiwan and received a First Class Certificate from the American Orchid Society. It was derived from crossing *P.* Snow Daffodil (a tetraploid white hybrid) with a diploid F1 hybrid with brown spots and yellow background color, *P.* Mambo (*P. amboinensis* × *P. mannii*). Nonetheless, there is no colchicine-treated plants of *P.* Golden Emperor 'Sweet' available to rescue its low fertility.

One more critical case in breeding yellow *Phalaenopsis* is breeding line of *P.* Taipei Gold. In 1980s, *P.* Taipei Gold was produced by crossing *P.* Glady's Read 'Snow Queen' (a tetraploid white hybrid) with a diploid species with dark bronze color from Sulawesi, *P. venosa*. Before the blooming of first plant, it was hard to imagine what color will be contributed by *P. venosa*. Breeders were surprised that many seedlings of *P.* Taipei Gold had clear yellow flowers, such as *P.* Taipei Gold 'STM'

Fig. 5.1 Flowers of *Phalaenopsis* yellow cultivars. (**a**) *P.* Golden Sands 'Canary' (**b**) The hexaploid *P.* Golden Sands 'Canary' (**c**) *P.* Taipei Gold 'STM' (**d**) *P.* Taipei Gold 'Gold Star' (**e**) *P.* Golden Emperor 'Sweet' (**f**) *P.* Fuller's Sunset (**g**) *P.* Fortune Saltzman 'Maple Bridge' (**h**) *P.* Joy Spring Canary 'Taipei' Scale bars = 1 cm

(Fig. 5.1c). But hardly surprising, most seedlings of *P.* Taipei Gold are triploid and almost sterile (Lee et al. 2020). In 1990s, breeders in Taiwan found the fertile cultivar, *P.* Taipei Gold 'Gold Star' (Fig. 5.1d), which could be used as both pollen and pod parents. Afterward, *P.* Taipei Gold 'Gold Star' was used broadly to cross with many yellow cultivars and produced a number of

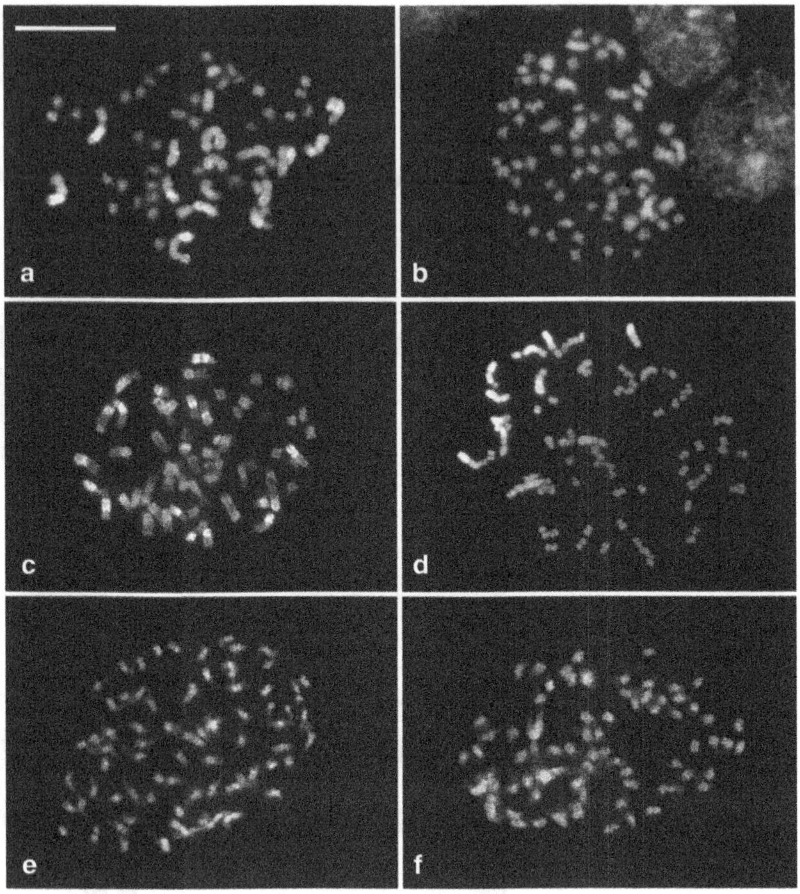

Fig. 5.2 Chromosomes of *Phalaenopsis* yellow cultivars. (**a**) *P.* Golden Emperor 'Sweet' (2n = 3x = 57) (**b**) *P.* Fortune Saltzman 'Maple Bridge' (2n = 4x − 6 = 70) (**c**) *P.* Joy Spring Canary 'Taipei'(2n = 3x = 57) (**d**) *P.* Fuller's Sunset (2n = 4x − 1 = 75) (**e**) *P.* Sin-Yaun Golden Beauty (2n = 4x − 2 = 74) (**f**) *P.* Fusheng's Golden Age (2n = 4x = 76) Scale bars = 10 μm

significant yellow cultivars, e.g. *P.* Brother Nugget (*P.* Taipei Gold × *P.* Brother Imp), *P.* Brother Stage (*P.* Taipei Gold × *P.* Golden Amboin), *P.* Sogo Lisa (*P.* Taipei Gold × *P.* Salu Spot). *P.* Taipei Gold 'Gold Star' with high fertility is owing to the tetraploidy, containing 38 large and asymmetrical chromosomes derived from the unreduced gametes of *P. venosa* (Lee et al. 2020).

Finally, a famous yellow breeding line, *P.* Deventeriana, is a F1 hybrid between a white flower species, *P. amabilis* and *P. amboinensis* possessing brown spots and yellow background color flowers. The cultivar, 'Treva' received an Award of Merit from the American Orchid Society, and it has been used extensively in the breeding of yellow and other novel colors *Phalaenopsis* since 1980s, such as *P.* Golden Gift (*P.* Deventeriana × *P.* Golden Buddha), *P.* Orchid World (*P.* Malibu Imp × *P.* Deventeriana), *P.* Sierra Gold (*P.* Deventeriana × *P.* Mambo) and *P.* Sweet Memory (*P.* Deventeriana × *P. violacea*) (Royal Horticultural Society 2019). According to the report by Aoyama (1993), *P.* Deventeriana 'Treva' is a tetraploid plant, containing 38 large and asymmetric chromosomes from *P. amboinensis* and 38 small chromosomes from *P. amabilis*. Hence, it is possible that the breeder used a tetraploid *P. amabilis* to cross with the diploid *P. amboinensis*, with the occurrence of unreduced gamete in *P. amboinensis*. The high fertility of *P.* Deventeriana 'Treva' made it an important breeding stock among the breeders in Taiwan and USA. A number of famous progenies in yellow *Phalaenopsis* breeding were produced in 1990s, including *P.* Brother Knight (*P.* Brother Imp × *P.* Deventeriana), *P.* Brother Lawrence (*P.* Taipei

Gold × *P.* Deventeriana), *P.* Brother Paradise (*P.* Deventeriana × *P.* Brother Carol) and *P.* Orchidview Bellringer (*P.* Deventeriana × *P.* Golden Bells). In the cases of *P.* Taipei Gold 'Gold Star' and *P.* Deventeriana 'Treva', the production of unreduced gametes in the *Poly chilos* species may play an important role in the breeding of yellow *Phalaenopsis*.

It is worthy to note that the tetraploid *P.* Deventeriana 'Treva' crossed with diploid parents may result in the formation of triploid progenies, such as *P.* Orchid World 'Bonnie Vasquez' (*P.* Malibu Imp × *P.* Deventeriana), *P.* Sierra Gold 'Suzanne' (*P.* Deventeriana × *P.* Mambo) and *P.* Sweet Memory (*P.* Deventeriana × *P. violacea*). They usually produce a few seedlings within a capsule. In our investigations, *P.* Joy Spring Canary 'Taipei' (*P.* Buena Jewel × *P.* Yungho Gelb Canary) is a triploid plant (Fig. 5.1h). In the chromosome composition of *P.* Joy Spring Canary 'Taipei', there are several large chromosomes (Fig. 5.2c) because of high proportion of species in subgenus *Polychilos*, i.e. *P. amboinensis*, *P. micholitzii* and *P. violacea* existed in its pedigree. This is the case of the occurrence of unreduced gametes in one of diploid parents, and thus resulted in the formation of triploid plant. *P.* Joy Spring Canary 'Taipei' has a relatively large and thick flowers as compared with its parents but produces merely a few seedlings within a capsule.

There were also other breeding lines of yellow *Phalaenopsis* worthy to be mentioned here, such as *P.* Autumn Sun (*P.* Prospector's Dream × *P.* Autumn Leaves) and *P.* Chia Lin (*P.* James Hall × *P.* Johanna). In 1991, *P.* Autumn Sun was bred by Herbert Hager and registered by Orchid Zone in California. Afterward, a few cultivars of *P.* Autumn Sun with clear bright yellow color were introduced to Taiwan, and breeders used broadly in breeding programs, some renowned hybrids such as *P.* Sogo Manager (*P.* Brother Lawrence × *P.* Autumn Sun), *P.* Sogo Medal (*P.* Autumn Sun × *P.* Taipei Gold) and *P.* Yu Pin Natsume (*P.* Autumn Sun × *P.* Haur Jin Diamond) have substantial influence in breeding yellow *Phalaenopsis*. Although we do not have plant materials of *P.* Autumn Sun, it is suggested that *P.* Autumn Sun should be a tetraploid or near tetraploid hybrid because of high fertility of its progenies mentioned above.

P. Chia Lin was bred and registered in 1982, but it was frequently used in breeding programs in 2000s. Some important progenies of *P.* Chia Lin, such as *P.* Chian Xen Queen (*P.* Chia Lin × *P.* Mount Beauty) and *P.* Sunrise Star (*P.* Chia Lin × *P.* Tinny Honey) have a substantial influence in breeding yellow *Phalaenopsis* with red lip cultivars, for example *P.* Fuller's Sunset (*P.* Taisuco Date × *P.* Chian Xen Queen) (Fig. 5.1f) and *P.* Fusheng's Golden Age (*P.* Fong-Tien's Yellow Butterfly × *P.* Chian Xen Queen). The yellow pigment of *P.* Chia Lin came from its grandparent, *P.* Barbara Moler [*P.* James Hall (*P.* Red Lip × *P.* Barbara Moler)], and it can be finally traced back to a diploid F1 hybrid, *P.* Yardstick (*P. fasciata* and *P. hieroglyphica*). It is also worthy to note that a famous orange color *Phalaenopsis* cultivar, *P.* Sin-Yaun Golden Beauty came from the cross between *P.* James Hall and *P.* Tsuei You Beauty. Although we do not have plant materials of *P.* Chia Lin and *P.* Barbara Moler, it is possible that these two hybrids are tetraploid plants due to the occurrence of unreduced gametes in *P.* Yardstick. From the data of cytological analyses, *P.* Fuller's Sunset is a hypotetraploid plant ($2n = 4x - 1 = 75$) (Fig. 5.2d); *P.* Sin-Yaun Golden Beauty is a hypotetraploid plant ($2n = 4x - 2 = 74$) (Fig. 5.2e); *P.* Fusheng's Golden Age is tetraploid plant ($2n = 4x = 76$) (Fig. 5.2f). They possess different number of large chromosomes in their chromosome compositions.

5.3 Prospective

In the yellow *Phalaenopsis* cultivars with standard-type flower size and flower arrangement, these cultivars carry different numbers of large chromosomes from section *Polychilos* species in their chromosomal compositions. Differential introgression of chromosomes from section *Polychilos* species may affect the color intensity of yellow *Phalaenopsis* cultivars. Since genome references of *P. equestris* and *P. aphrodite* are available (Cai et al. 2015; Chao et al. 2018), it would be feasible to map the chromosome carrying the genes involved in metabolism of yellow pigment, e.g. chalcone synthase in the future. The application of fluorescent in situ hybridization (FISH) and genomic in situ hybridization (GISH) would be useful to identify specific chromosomesor segments of interest in *Phalaenopsis* hybrids.

Acknowledgements The authors are grateful for the support from Institute of Plant and Microbial Biology, Academia Sinica, Taiwan to Mei-Chu Chung, and by the Ministry of Science and Technology, Taiwan [NSC 102-2313-B-178-001-MY3] and National Museum of Natural Science, Taiwan to Yung-I Lee.

References

Aoyama M (1993) Polyploidy in *Phalaenopsis* hybrids. Bull. Hiroshima Prefect. Agric Res Cent 57:55–62

Aoyama M (2010) Chromosome research for orchid breeding. Acta Hortic 878:125–132

Arends JC (1970) Cytological observations on genome homology in eight interspecific hybrids of *Phalaenopsis*. Genetica 41:88–100

Bolanos-Villegas P, Chin SW, Chen FC (2008) Meiotic chromosome behavior and capsule setting in *Doritaenopsis* hybrids. J Amer Soc Hortic Sci 133:107–116

Cai J, Liu X, Vanneste K et al (2015) The genome sequence of the orchid *Phalaenopsis equestris*. Nat Genet 47:65–72

Chao YT, Chen WC, Chen CY et al (2018) Chromosome-level assembly, genetic and physical mapping of *Phalaenopsis aphrodite* genome provides new insights into species adaptation and resources for orchid breeding. Plant Biotechnol J 16:2027–2041

Chuang HT, Hsu ST, Shen TM (2008) Breeding barriers in yellow *Phalaenopsis* orchids. J Taiwan Soc Hortic Sci 54:59–66

Freed H (1980a) *Phalaenopsis amboinensis*. Amer Orchid Soc Bull 49:468

Freed H (1980b) The fabulous *Phalaenopsis fasciata*. Amer Orchid Soc Bull 49:1099

Freed H (1981a) The versatile *Phalaenopsis lueddemanniana*—part 1. Amer Orchid Soc Bull 50:1077

Freed H (1981b) The versatile *Phalaenopsis lueddemanniana*—part 2. Amer Orchid Soc Bull 50:1325

Griesbach RJ (1985) Polyploidy in *Phalaenopsis* orchid improvement. J Hered 76:74–75

Griesbach RJ (2002) Development of *Phalaenopsis* orchids for the mass-market. In: Janick J, Whipkey A (eds) Trends in New Crops and New Uses. ASHS Press, Alexandra, VA, pp 458–465

Lee YI, Chung MC, Kuo HC, Wang CN, Lee YC, Lin CY, Jiang H, Yeh CH (2017) The evolution of genome size and distinct distribution patterns of rDNA in *Phalaenopsis* (Orchidaceae). BotJ Linn Soc 185:65–80

Lee YI, Tseng YF, Lee YC, Chung MC (2020) Chromosome constitution and nuclear DNA content of *Phalaenopsis* hybrids. Sci Hortic 262:27. https://doi.org/10.1016/j.scienta.2019.109089

Royal Horticultural Society (2019) Royal Horticultural Society orchid hybrid registration database information system. https://apps.rhs.org.uk/horticulturaldatabase/orchidregister/orchidresults.asp/

Tang CY, Chen WH (2007) Breeding and development of new varieties in *Phalaenopsis*. In: Chen WH, Chen HH (eds) Orchid Biotechnology. World Sci Pub, Singapore, pp 1–22

United States Department of Agriculture (2019) Floriculture crops 2019 summary. United States Department of Agriculture, Washington DC

Van Tuyl JM, Lim KB (2003) Interspecific hybridisation and polyploidisation as tools in ornamental plant breeding. Acta Hortic 612:13–22

Vaughn L, Vaughn V (1973) An account of moth orchids—the ascendancy of white *Phalaenopsis*. Amer Orchid Soc Bull 42:231–237

Regulation of Flowering in Orchids

6

Jian-Zhi Huang, Pablo Bolaños-Villegas, and Fure-Chyi Chen

Abstract

Orchids constitute the largest families within the flowering plants; they are one of the most highly evolved groups in the angiosperms; and because of their flashy flowers many species from several genera are very prominent horticultural commodities. Environmental factors such as temperature and daylength influence flowering; however, other factors play a major role such as the developmental stage, plant hormonal regulation and genetic factors. Research on model plants such as the thale cress (*Arabidopsis thaliana*) and rice (*Oryza sativa*) has allowed a broad understanding of the genetics and the molecular regulation of flowering in most plants. However, until very recently the specifics of floral regulation in orchids had remained quite obscure. In this review we have examined very recent work concerning flowering of orchids with a focus on the interplay between development, environmental cues, plant hormones, and gene networks. We have highlighted several examples of successful manipulation of flowering through biotechnology that may be very relevant for researchers and growers elsewhere.

6.1 Introduction

For sexual reproduction to succeed, timing is essential. Flowering in orchids has to occur at the right time for the production of large capsules filled with healthy seeds. A plant determines to flower based on the integration of environmental stimuli such as light conditions, daylength, temperature, drought conditions and nutrient availability; endogenous factors such as developmental maturity, hormonal balance and age; and the effect of agricultural plant growth regulators, pests and diseases (Cho et al. 2017; He and Amasino 2005; Song et al. 2013). In *Arabidopsis thaliana*, genetic dissection of flowering pathways suggests the existence of six key flowering pathways, namely the photoperiod pathway, the autonomous pathway, the aging pathway, the gibberellin (GA) pathway, and the vernalization and environment temperature

J.-Z. Huang · F.-C. Chen (✉)
Department of Plant Industry, National Pingtung University of Science and Technology, Pingtung 91201, Taiwan
e-mail: fchen@mail.npust.edu.tw

J.-Z. Huang
Department of Agricultural Chemistry, College of Bioresources and Agriculture, National Taiwan University, Taipei 10617, Taiwan

P. Bolaños-Villegas
Fabio Baudrit Agricultural Research Station, University of Costa Rica, Alajuela 20101, Costa Rica

P. Bolaños-Villegas
Jardín Botánico Lankester, Universidad de Costa Rica, San José, Costa Rica

F.-C. Chen and S.-W. Chin (eds.), *The Orchid Genome*, Compendium of Plant Genomes,
https://doi.org/10.1007/978-3-030-66826-6_6

pathway (Andrés and Coupland 2012; Mouradov et al. 2002). At least 306 genes have been shown to regulate the flowering time and their respective profiles have been compiled by the Flowering Interactive Database (FLOR-ID) (Andrés and Coupland 2012; Bouché et al. 2016; Johansson and Staiger 2014; Simpson and Dean 2002). It is believed that ultimately all these pathways converge to two major regulators and integrators, the *FLOWERING LOCUS T* (*FT*) and the *SUPPRESSOR OF OVEREXPRESSION OF CO1* (*SOC1*), which in turn target genes that determine cell fate and identity in the floral meristem, including *APETALA1* (*AP1*), *LEAFY* (*LFY*) and *CAULIFLOWER* (*CAL*) (Abe et al. 2005; Corbesier et al. 2007; Kieffer and Davies 2001; Lee and Lee 2010; Lee et al. 2008; Lohmann et al. 2001). The *FT* gene is believed to be the regulator of florigen expression, a flowering-promoting protein that is conserved across higher plants (Ahn et al. 2006; Corbesier et al. 2007; Kobayashi et al. 1999).

Flowering is not a trivial issue for orchid growers. Every year millions of orchids around the world are sold either as potted plants or their flower spikes are harvested and sold as cut flowers. Orchids are also sold for the ornamental value of their foliage as in the case of Asian genus *Cymbidium*, known as "*lan*" in Chinese (Koehn 1952), which combines beautiful leaf variegation patterns and beautiful flowers (Gao et al. 2020). It is therefore not surprising that there has been significant interest in developing methods to promote precisely scheduled flowering of orchids for trade. Literature related to orchid flowering does exist, but it is often lacking in details (Goh 1979; Kardailsky et al. 1999; Rotor 1952), and while research from *Arabidopsis* and rice has allowed to sketch the architecture of regulatory networks that control flower induction and floral differentiation, this knowledge may not apply to most other crops. Therefore, we felt that it is necessary to summarize recent work in orchids that allow to place together the pieces of a puzzle that show the regulatory mechanisms of flowering in commercial orchids.

6.2 Flower Induction and Differentiation of Orchids

The processes of floral evocation and induction are thought to be multifactorial and sequential (Bernier 1988). The actual process of flowering can be further divided into two substages: flower induction and flower organ development. After flower induction, the flower bud/inflorescence begins to develop but its ensuing growth rate will be determined by the availability of photosynthetic assimilates until flowers are fully developed. The source of these assimilates will be the leaves, and in most plants the upper leaves preferably supply photosynthetic nutrients to the flower bud and then supply a remainder to the stem (Mor and Halevy 1979). The orchid inflorescence emerges from either terminal or axillary meristems. The phase change from apical to floral (vegetative to reproductive) is a critical developmental step in orchids, and it is thought to be regulated by identifiable environmental and endogenous factors. Orchids are categorized into two basic growth types determined by the position and the number of fully differentiated bud primordia which later develop into flowering shoots (Rotor 1952). The first category is that of plants that feature a pseudobulb with a single terminal bud primordium such as in *Cattleya* (Fig. 6.1a) and *Paphiopedilum* (Fig. 6.1b). Only the apical bud primordium in a newly formed pseudobulb is able to give rise to a fully developed inflorescence. The equitant *Oncidium* (now *Tolumnia*) develops only one bud primordium per pseudobulb, and each plant will produce a single inflorescence spike which is capable of producing flower outshoots from dormant buds on the rachis (Fig. 6.1c). The second category consists of orchid genera with multiple bud primordia that are located in the leaf axils, such as in *Phalaenopsis* (Fig. 6.1d), *Vanda* (Fig. 6.1e), *Dendrobium* (both *nobile* type and *phalaenopsis* type) (Fig. 6.1f, g), *Cymbidium* (Fig. 6.1h), *Eria* and in *Oncidesa* (Fig. 6.1i). Special adaptations do exist, though, for instance, in the case of *Cymbidium*, each pseudobulb will flower just once, but a plant may simultaneously grow several pseudobulbs that bloom at once (Fig. 6.1h). In *Dendrobium*, leaf axils are the site where bud primordia form. *Dendrobium phalaenopsis* and its hybrids are a special case as well; these plants grow inflorescences on pseudobulbs both new and old, but a young pseudobulb that flowers for the very first time will produce a terminal inflorescence from the axil situated at the uppermost leaf. As the plant ages that pseudobulb will produce several new subterminal inflorescences, along with terminal inflorescences that the plant will grow on new pseudobulbs (Fig. 6.1g). In *Dendrobium* of the *nobile* type, plants respond to cold weather and bloom simultaneously at every node on old and mature pseudobulbs after a dry, resting stage (Fig. 6.1e). In *Oncidium*, plants will grow on green pseudobulbs from which a pair of long and narrow leaves emerge at the top, while a pair of short leaves grow at the base (Rittershausen and Rittershausen 2003; Blanchard and Runkle 2008b). As in the case of other sympodial orchids, the flower primordia of *Oncidesa* (Syn. *Oncidium)* (Fig. 6.1i) start to develop at the base of mature pseudobulb in the axil of the leaf sheath (Goh et al. 1982a, b).

Monopodial orchids have a growth habit that sets them apart. These orchids develop a main stem that grows upward, from which new leaves grow at the apex. Flowering then usually occurs at a fairly constant distance several leaf axis below the apical meristem. Two very famous monopodial genera from Asia are *Phalaenopsis* and *Vanda*, in which plants typically feature at least two bud primordia at each leaf axil. These primordia remain quiescent until stimuli trigger either one of them or both to develop into flower spikes or into asexual young offshoots referred to as 'keikis' (Abraham and Vatsala 1981). In addition, as a plant matures, increasing numbers of inflorescence spikes may be produced during each blooming episode. Environmental cues that signal the initiation of flowering are received by receptive primordium that will start to develop spikes and flowers. However, nearby primordia will remain in a quiescent state unless the apical stem primordium is damaged or destroyed. Then, the formerly quiescent primordium will develop into a keiki. Interestingly, old flower spikes that are not cut or removed may retain the ability to form multiple keikis, and this is believed to be an evolutionary strategy for survival. One notable example was first reported more than a century ago (Williams and Williams 1894), when plants in the species *Phalaenopsis lueddemanniana* were described for their ease at growing keikis on flower spikes. This trait is quite common in species that bloom in the summer, such as *Phalaenopsis pulchra* (Fig. 6.2a) and *Phalaenopsis equestris* (Fig. 6.2b), however hybrids often lose the trait. In other vandaceous species, damage or removal of the apical stem induces formation of asexual offshoots at axial buds down below (Teoh 2005).

6.3 Factors that Control Flower Induction

In orchids the time to flower is thought to be determined by sexual juvenility, photoperiod and cues that control vernalization. Ontogeny and seasonal cues are also believed to be important by growers and researchers (Wang et al. 2019).

6.4 Juvenility and Control of Flowering in Orchids

Juvenility in plants is defined as an early stage of growth in which individuals are insensitive to environmental cues that normally trigger flowering (Bernier et al. 1981; Damann and Lyons 1993; Pillitteri et al. 2004).

The length of the juvenile stage varies considerably among plant species. In some it lasts for only a few days (Cumming 1959; Friend 1968), whereas in some woody species it may last for up to 30–40 years (Hackett 1986). The transition from juvenility to a reproductive stage is usually brought about by an environmental signal such a vernalization stimulus, by a change in photoperiod or by both (Poethig 2010). In the

Fig. 6.1 Flowering characteristics in selected orchids. **a** *Laeliocatanthe* Orange Fantasy; **b** *Paphiopedilum* Maudiae hybrid; **c** *Tolumnia* sp.; **d** *Phalaenopsis philippinensis*; **e** *Vanda* hybrid; **f** Nobile type- *Dendrobium* Chanthaboon Sunrise; **g** Phalaenopsis type- *Dendrobium* hybrid; **h** *Cymbidium kanran* derived hybrid; **i** *Oncidesa* Gower Ramsey 'Honey Angel'

case of orchids it has been estimated that the average duration of the juvenile phase is 2–3 years depending on the species or genotypes, but extreme cases have been reported in *Aranda* Lucy Laycock (13 years) (Wee 1971), *Vandachnis* Scarlet Runner (11–12 years), *Vanda* John Warne (10 years) *Renantanda* Joan Mah (9 years), *Arachnopsis* Eric Holttum (7 years and a half) and *Cymbidium* Faridah Hashim (5 years) (Goh et al. 1982a, b; Wee 1971) (Table 6.1). Fortunately, plant breeding in Taiwan and other South East Asian locations has shortened the time to flowering to only 12–36 months after germination. In these plants the period may be reduced further by careful control of irrigation frequency and duration, the ratio of fertilizers and temperature in the greenhouse.

Fig. 6.2 **a** Keikis in *Phalaenopsis pulchra,* an early-summer flowering species; **b** A keiki bears flowers in a miniature *Phalaenopsis equestris* mother plant

Table 6.1 Time from seed sowing to flowering in some orchid hybrids

Orchid hybrid	Juvenile period (years–months–days)
Arachnopsis Eric Holttum	7–5–2
Aranda Hilda Galistan	4–8– 8
Aranda Lucy Laycock	13–3–0
Aranda Wendy Scott	7–10–6
Aranthera Anne Block	5–10–5
Aranthera Beatrice Ng	6–1–3
Burkillara Henry	5–9–22
Cymbidium Faridah Hashim	5–0–20
Dendrobium Sarie Marijs	3–4–10
Dendrobium Lin Yoke Ching	8–2–12
Holttumara Cochinea	1 8–0–24
Cattlianthe Cheah Chuan Keat	6–7–14
Paphiopedilum Shireen	8–5–0
Mayara Joan Mah	9–0–0
Renantandra Storiata	9–3–1
Spathoglottis Penang Beauty	2–11–13
Vanda Miss Joaquim	3–1–5
Vanda Ruby Prince	3–4–16
Vanda Tan Chin Tuan	8–4–2
Vanda John Warne	10–2–3
Vandachnis Scarlet Runner	11–10–11

Adapted from Wee (1971) and Hew and Yong (2004) with updated grex genus after Royal Horticultural Society

6.5 The Effect of Temperature in Orchid Flowering

Temperature has a strong influence on the flowering behavior of plants (Craufurd and Wheeler 2009; Jagadish et al. 2016). In plants adapted to temperate climates, including *Arabidopsis thaliana* (Imaizumi and Kay 2006; Jaeger et al. 2006) and cereal crops such as barley (*Hordeum vulgare*) and wheat (*Triticum aestivum*) (Distelfeld et al. 2009; Trevaskis et al. 2007), the transition from the juvenile stage to a reproductive phase can be precipitated by long exposure to cold. This process is known as vernalization. Several orchid genera require an obligate episode of vernalization after shoot maturation is complete to flower, such as in *Phalaenopsis* and *nobile*-type *Dendrobium*. During the cold treatment period, a grower may manage the temperature, the time of initiation and its duration, in addition to changing the day/night temperature (Yen 2008; Yen et al. 2008; Yoneda et al. 1991). Temperature has been reported to be a determinant factor for the control of initiation of flower bud development in commercial orchid genera such as *Dendrobium* (Campos and Kerbauy 2004; Rotor 1952), *Miltoniopsis* (Lopez and Runkle 2006), *Phalaenopsis* (Blanchard and Runkle 2006), *Zygopetalum* (Lopez et al. 2003), *Odontioda* (Blanchard and Runkle 2008b), *Cattleya* (Lopez and Runkle 2005; Rotor 1952) and *Cymbidium* (Kim et al. 2011; Yang et al. 2019). A common practice during the commercial cultivation of *Phalaenopsis* within greenhouses is that automated climate controls are set to maintain the temperature above 28 °C to keep a robust rate of vegetative growth while inhibiting the reproductive transition (Sakanishi et al. 1980; Chen et al. 1994; Newton and Runkle 2009). Then, once *Phalaenopsis* plants have developed large and healthy leaves, the night temperature is set to a cool condition (below 26 °C) in order to promote inflorescence formation, or spiking (Lee et al. 1987; Sakanishi et al. 1980; Yoneda et al. 1991; Wang 1995). Blanchard and Runkle (2004) reported a high flowering rate of 80% in hybrids *Phalaenopsis* Miva Smartissimo x Canberra '450' and *Phalaenopsis* Brother Goldsmith '720', when grown first at 20–25 °C for six weeks, and then switched to day/night temperatures of 23/17 °C. Variation in the response to temperature has been reported, for instance in *Phalaenopsis* exposure to temperatures above 28°C triggers the formation of keikis instead of floral buds, and the remaining flower buds may be aborted (Sakanishi et al. 1980). Newton and Runkle (2009) reported that hybrids *Phalaenopsis* Miva Smartissimo x Canberra 'Mosella', *Phal.* Brother Pink Mask x Brother Success 'Explosion', *Phal.* Baldan's Kaleidoscope 'Golden Treasure' and *Phalaenopsis* 'Newberry Parfait' may require 8–12 h at a temperature of 29 °C to completely avert spiking. In the case of beautiful species *Phalaenopsis schilleriana*, a regime of three weeks of high night temperatures followed by a drop below 21 °C is necessary for induction of flowering (De Vries 1953). Rotor and Withner (1959) reported that for the commercially important species *Phalaenopsis amabilis*, it is possible to induce continuous flowering the whole year when grown under a short-day photoperiod and nighttime temperature below 20 °C. An opposite regime restricts flowering to occur only once per year.

The genus *Dendrobium* is one of the largest in the Orchidaceae family and comprises more than 1,200 species (Eigeldinger and Murphy 1972; Baker and Baker 1996). In most species of this large genus, steady vegetative proliferation and growth are stimulated by temperatures of 24–30 °C (Leonhardt 2000). A low temperature of about 13 °C induces flowering in species such as *Dendrobium phalaenopsis* Fitz var. *statterianum* Hort. ex Sander (Rotor 1952). In species *Dendrobium crumenatum* Swartz, natural growth conditions feature a long period of relative cold that induces initial flower bud development. The flower buds remain dormant until a further drop in temperature of about 5 °C occurs, and development resumes. This drop in temperature is linked to rainstorms in its habitat in South East Asia (Goh et al. 1982a, b; Arditti 1979; Holttum 1949). This pattern of obligate vernalization and flowering after an initial period of pseudobulb maturation is common in all dendrobiums of the *nobile*-type. For instance, Rotor (1952) reported that exposure to a cold temperature regime of about 18 °C for four months caused a delay in flowering in species *Dendrobium nobile* Lindley, but plants exposed to temperatures below 13 °C were induced to flower. In hybrid *Dendrobium* Snow Flake 'Red Star', relatively high daytime temperatures inhibit flower bud development, reduce sharply the number of flower buds and redirect development into the formation of keikis (Sinoda et al. 1988). Yen (2008) determined that in *Dendrobium* Sea Mary 'Snow King' the mandatory cooling period is apparently saturated by a regime of 13 °C for 3 weeks, and without any exposure to cool temperatures these plants totally fail to bloom and instead either abort buds or produce keikis.

Cymbidium, the slender and delicate "*lan*" orchid immortalized in traditional Chinese painting (Koehn 1952), is a rather small genus of approximately 70 species of tropical and subtropical origin (Pridgeon 2000). The genus is primarily distributed in Asia, although several species are native to northern Australia. These orchid species require long days and daytime and nighttime temperatures of 30 and 25 °C for quick vegetative development and proper maturation of pseudobulbs. Photoperiod does not seem to influence flower bud initiation in *Cymbidium*, but low temperatures are absolutely required (Ichihashi et al. 1997; Rotor 1952; Pridgeon 2000). Went (1957) found that in species *Cymbidium giganteum* var. *lowianum* Rchb. f., fluctuations in daytime to nighttime temperatures of either 26/14 °C or 23/14 °C, do not induce initiation of flowering; nonetheless, diurnal/nocturnal temperatures of 26/7 °C or 26/10 °C trigger flowering, with an optimal yield at 20/10 °C and 20/14 °C. Lopez and Runkle (2005) confirmed that in *Cymbidium* there is a requirement for separate daytime and nighttime temperatures in order to induce optimal flowering. Powell et al. (1988) reported that for the specific case of *Cymbidium* Astronaut 'Rajah' flowering requires a 14 h photoperiod and daytime/nighttime temperatures of 26/12 °C. In small and miniature *Cymbidiums* such as *Cymbidium pumilum* and *Cymbidium* Sazanami 'Haru-no-umi' temperatures above 25 °C are detrimental for cultivation and may cause early cessation of flower bud growth and flower abortion (Ohno 1991; Ichihashi et al. 1997).

The genera *Cattleya/Guarianthe* are native to tropical and neotropical America and comprises approximately 50 species (Lopez and Runkle 2004a; Dressler and Higgins 2003). Flowering in these orchids is stimulated by short days and relatively low temperatures during the first months of the year, but flower buds enlarge and open until a few months later (Lopez and Runkle 2004a). Environmental requirements for flowering may vary, though. In *Cattleya warscewiczii*, *C. gaskelliana* and *C. mossiae*, flower induction takes place only at 13 °C for a 9 h photoperiod, while a 16 h photoperiod inhibits flowering (Rotor 1952; Rotor and Withner 1959). In *C. warscewiczii* flowering can be accelerated by a regime of low temperature and short photoperiod that promotes pseudobulb development for subsequent rapid flowering (Rotor 1952).

The *Odontoglossum* alliance includes species that are native to the highlands of the American tropics and are evolutionarily part of the *Oncidiinae* tribe (Dalström and Higgins 2016). *Odontioda* is an intergeneric genus (*Cochlioda* × *Odontoglossum*) that requires relatively cool temperatures and is immensely popular with hobbyists (Blanchard and Runkle 2008b). Cool temperatures (14–17 °C) are responsible for eliciting over 90% of flowering in plants from *Odontioda* Lovely Penguin 'Emperor' and *Odontioda* George McMahon 'Fortuna' when compared to 20–23 °C (Blanchard and Runkle 2005, 2008b). *Miltoniopsis*, or pansy orchid, is another genus of enormous commercial importance related to the *Odontoglossum* alliance (Ritterhausen and Ritterhausen 2003). This is a genus of lithophytic and epiphytic plants that comprises only five species naturally found in humid lowlands to the wet cloud forest (610–2100 m), in a very long latitude range from Costa Rica to Perú (Baker and Baker 1993; Morrison 2000; Lopez and Runkle 2005). In the *Miltoniopsis* Augres 'Trinity' hybrid efficient flowering is induced when plants are cultivated under a short-day regime and then vernalized at a temperature of 11–14 °C (Lopez and Runkle 2006). *Zygopetalum* is another flashy genus from this alliance. This is a genus of sympodial, lithophytic and epiphytic plants from South America that groups about 14-20 species (Rittershausen and Rittershausen 2003). In hybrid *Zygopetalum* Redvale 'Fire Kiss', it is possible to induce quick and uniform flowering when cultivated on a 9 h photoperiod for eight weeks and then vernalized at a temperature of 11–14 °C for other 8 weeks (Lopez et al. 2003; Lopez and Runkle 2004a).

6.6 Flowering of Orchids as a Response to Photoperiod

Daylength exerts profound effects on many developmental functions in plants. The existence of this environmental factor was first proposed in 1918 by Klebs. The mechanism underlying the daylength (photoperiodic) response in plant development was first described in detail in 1920 (Garner and Allard 1920).

The concept circadian clock was first proposed by Bünning in 1936 as an endogenous timing mechanism for a duration of 24 h (Bünning 1936; Lumsden 2002). Since then, photoperiodism has been a subject of intensive research in plant science. The response of plants to the length of daylight seems to fall into three basic types: short-day plants (SDP), long day plants (LDP) and day neutral plants (DNP) responses. For instance, model plants *Arabidopsis thaliana* and *Oryza sativa* are LDP and SDP types, respectively, and their responses have been dissected at the molecular level (Hayama and Coupland 2004; Yanovsky and Kay 2003; Corbesier and Coupland 2005; Searle and Coupland 2004; Imaizumi and Kay 2006). Orchids fall into all three categories depending on the genus (either SDP, LDP or DNP); moreover, the impact of the photoperiod on the regulation of vegetative growth varies depending on whether the plant is a species or a hybrid (Lopez and Runkle 2004b).

Evidence suggests that floral initiation in orchids is influenced both by the photoperiod and the genotype. A short day for flower initiation is required for the flowering in orchids such as in *Zygopetalum* (Lopez et al. 2003). However, in *Phalaenopsis*, *Cymbidium*, *Vanda* and several *Dendrobium* hybrids, there is either total insensitivity to the photoperiod or no apparent response is ever observed, perhaps due to the prevalence of the response to temperature; therefore these plants were treated as DNP types (Goh 1985; Lopez and Runkle 2005). Similarly, experiments conducted by Sheehan (1983) *Dendrobium* Jacquelyn Thomas and *Dendrobium* Lady Hay revealed that both are able to flower throughout the year, suggesting that these plants lost sensitivity to daylength. A few studies found that in *Phalaenopsis* the emergence of the flower spike as well as its length is positively influenced by short-day regimes, while long days stimulate vegetative proliferation and the growth of keikis (De Vries 1953; Rotor 1952; Griesbach 1985; Yoneda et al. 1991). It is also known that in order to flower, orchids native to relatively temperate habitats have a requirement for short days and low temperatures, such as in a few species from *Cattleya*, *Dendrobium* and *Phalaenopsis* (Sheehan 1983). However, most species from *Phalaenopsis* do not show any apparent photosensitivity (Lopez and Runkle 2005). In *Cattleya*, flowering in species and hybrids is efficiently induced by cultivation at 13 °C with a daylight regime of 9 h; the opposite happens when plants are grown at the same temperature but are exposed to 16 h of light (Rotor 1952, Rotor and Withner 1959). This genus is remarkable in the sense that contains species from both the LDP and SDP types of flowering (Lopez and Runkle 2004a, Rotor 1952, Rotor and Withner 1959). As an example, species such as *Cattleya labiata*, *C. trianaei* and *C. mossiae*, are positively stimulated to flower by short days, while long days suppress flowering; in contrast, *C. granulosa*, *C. dowiana* and *C. intermedia* require long days to initiate development of flower buds (Batchelor 2011). In Asian monopodial genus *Vanda* Murashige et al. (1967) found that *Vanda* x Miss Joaquim grows more flowers in full sunlight. Goh and Wan (1973) also reported that in *Vanda* x Miss Joaquim the induction of flowering is more robust if plants are grown at a 10 h photoperiod; and that short exposure times reduce the number of flower spikes developed.

In orchids the response to photoperiod can be modified by temperature. In fact, in most cultivated orchids plants are no longer sensitive to daylength but respond well to low temperature. For example, short-day cultivation of *Miltoniopsis* at 23 °C with a shift to 11–14 °C induces strong flowering, but the photoperiod did not show any significant effect suggesting that the environmental temperature is a decisive factor when it comes to inducing flowering in *Miltoniopsis* (Lopez and Runkle 2006). In *Dendrobium nobile*, flowering seems to be affected mainly by temperature. In temperatures below 13 °C flowers are produced regardless of daylength. Longer days are reported to hasten blooming by 1–4 weeks. In contrast, *Dendrobium phalaenopsis* requires short daylengths at both 13 and 18 °C for flowering (Rotor 1952, Rotor and Withner 1959). As mentioned earlier, in *Zygopetalum* cool temperatures (11–14 °C) and a short photoperiod are required for proper stimulation of flowering (Lopez et al. 2003, 2004). The twig epiphyte *Erycina pusilla* is a very special case since it does not show a sharp response to daylength, instead an increase in the duration of daytime also increases the number of floral spikes per plant, thus it is believed to be a quantitative long-day plant (Vaz et al. 2004). Thus, the impact of photoperiod in the promotion of orchid flowering seems to vary among species and hybrids.

Tropical orchids whose natural habitat is close to the Equator are considered to be less responsive to minor differences in daylength than those that hail from temperate regions (Sanford 1974). The reason is that in tropical latitudes there is little or no seasonal variation in daylength. Thus, it is reasonable to anticipate that changes in photoperiod may have limited effects on the flowering of truly tropical orchids.

6.7 Effects of Phytohormones on Orchid Flowering

Many physiological activities during plant growth and development are controlled by phytohormones (Weyers and Paterson 2001; Gray 2004). The phytohormones that are believed to play a major role in plant flowering are the auxins (IAA), the cytokinins (CKs), the gibberellins (GAs), the abscisic acid (ABA), the brassinosteroids (BRs), ethylene, salicylic acid (SA), jasmonic acid (JA) and the recently identified strigolactone (SL) (Zwanenburg and Blanco-Ania 2018). Langridge (1957) first reported that the application of exogenous GAs could accelerate development in *Arabidopsis*, suggesting that GAs played an essential role in inducing flowering under non-inductive growth conditions. Moreover, it was shown that the *Arabidopsis gibberellin deficient1* (*ga1*) mutant is unable to synthesize GAs and is uncapable to flower under short-day conditions (Wilson et al. 1992; Chandler and Dean 1994; Xu et al. 1997). Cytokinins also play a key role as regulators of inflorescence architecture through control of meristem activity, which is linked to meristem identity (Kyozuka 2007; Werner and Schmülling 2009; Perilli et al. 2010). In *Arabidopsis* and rice, the endogenous level of cytokinins directly influences meristematic activity and the architectural complexity of inflorescences (Kieber and Schaller 2018). In species such as *Boronia megastigma*, *Chenopodium rubrum*, *Nicotiana tabacum*, *Tillandsia recurvata* and *Sinapis alba*, it has been reported that during floral transition and flower development the cytokinin content increases significantly within the apical meristem (Day et al. 1995; Macháčková et al. 1993; Dewitte et al. 1999; Jacqmard et al. 2002; Mercier and Endres 1999). It is believed that a key component of this process is the degradation of bioactive cytokinins by the cytokinin oxidase/dehydrogenase (CKX) (Mok and Mok 2001). In rice the *Gn1a/OsCKX2* gene encodes a cytokinin oxidase that when overexpressed leads to an increase in bioactive cytokinin levels, enhanced activity at floral meristems and higher grain yields (Ashikari et al. 2005). In the *Arabidopsis ckx3/ckx5* double mutant, there is an increase in cytokinin levels which causes a concomitant development of large flower meristems, large flowers, more flowers and more siliques per flower (Bartrina et al. 2011).

In orchids the use of phytohormones has been effective for the control of flowering (Goh and Yang 1978). It is firmly believed that auxin intervenes directly as a promoter of flower bud initiation (Nemhauser et al. 1998), although its specific mechanism may be different among different orchid species/hybrids. Application of 6-benzylaminopurine (BA) has been shown to promote an increase in the number of inflorescences per plant in several orchid species (Goh 1979; Blanchard and Runkle 2008a; Wu and Chang 2009; Nisha et al. 2012). However, in *Dendrobium*, it is BA that promotes flowering, while IAA antagonizes the promotive effect of BA. In hybrid *Aranda* Deborah, the way to induce flowering is by decapitation of the apical meristem, and evidence suggests that decapitation reduces the synthesis of IAA. Exogenous application of this hormone suppresses flowering after decapitation, but if the application is interrupted flowering resumes. A second application of exogenous IAA is not able to inhibit floral development again (Goh and Wan 1973). Interestingly, during in vitro cultivation of *Dendrobium* shoot tips, application of synthetic cytokinin thidiazuron (TDZ, a chemical defoliant) had a significant positive effect in the endogenous levels of IAA and in the induction of flower bud formation, hinting at a role for IAA in the shift from vegetative to reproductive development (de Melo Ferreira et al. 2006). It is not clear yet whether a flow of auxin is essential for bud initiation or whether this causes turnover and decrease of its own concentration. The auxin transport inhibitor triiodobenzoic acid (TIBA) interferes with the polar translocation of auxin and thus reduces its endogenous concentration (Audus 1972). TIBA inhibits the formation of inflorescences in *Cymbidium niveo-marginatum* Mak (Kostenyuk et al. 1999). However, in *Aranda* auxin antagonists such as TIBA, and maleic acid hydrazide (MH) were found to promote flower initiation, but the flower buds did not continue to development (Goh 1977).

Cytokinins applied exogenously have been shown to accelerate flowering (Bernier et al. 1993). In orchids several studies have examined how to induce flower spike formation with plant hormones, for instance, in *Phalaenopsis* a foliar spray with solution of either BA or BA + GA_3 after transfer of plants to cool highlands (500 m above sea level) did promote flowering, reduced the time for emergence of the flower spikes and reduced the overall time to full flowering by 13–25 days (Yoneda and Momose 1990). Application of 100–400 mg/L BA to potted plants of *Phalaenopsis* Happy Valentine increased the number of spike number, but decreased length of spikes at low temperature (23/18 °C day/night) (Kim et al. 2000). Similarly, spray applications of 200 or 400 mg/L BA also increased significantly the spike number in *Phalaenopsis* (Blanchard and Runkle 2008a). Application of BA significantly promotes an increase in the number of spikes and flowers in *Cymbidium* (Lee et al. 1998), *Dendrobium* Lousiae 'Dark', *Dendrobium* 'Nodoka', *Oncidium* (Higuchi and Sakai 1977; Goh 1985; Hew and Clifford 1993; Sakai et al. 2000), *Aranda* (Zaharah et al. 1986; Goh 1977), *Miltoniopsis* (Newton and Runkle 2015), *Aranthera*, *Holtumara*, and *Mokara* (Zaharah et al. 1986).

Gibberellins are phytohormones considered to play an essential role during the transition from vegetative to reproductive growth, and the development of flowers, seeds and fruits (Hedden and Sponsel 2015; Davière and Achard 2013). Matsumoto (2006) reported that in *Miltoniopsis* an application of 2.5 mM GA_3 during the first flowering season quickened the emergence of the inflorescence by 10.9–14.9 days for hybrid Bert Field 'Eileen' and by 48 days for hybrid Rouge 'Akatsuka'. In the case of *Miltoniopsis* Bert Field 'Eileen' the number of flower spikes increased from 2.2 to 3.0 per plant. Exogenous application of GA_3 in *Phalaenopsis hybrida* is effective in promoting flower development under high temperatures that are usually inhibitory, but it is also believed that endogenous GAs promote development of the flower spike and the flower buds under low temperatures. A possible explanation is that application of GA_3 promotes longitudinal cell proliferation in flower primordia (Chen et al. 1997, 1994). Su et al. (2001) showed significant evidence that on flower stems of *Phalaenopsis* the application of GA_3 under temperatures of 30/25 °C (day/night) increased the levels of bioactive GA_1 that stimulate floral induction, a phenomenon similar to flowering under cool temperatures. Thus, it is possible that in general the exogenous application of GA_3 and/or cool temperatures induce a decrease in the turnover rate of bioactive GA_1 to inactive GA_8 and stimulates flower induction and flower development (Su et al. 2001). In *Brassocattleya* Marcella Koss, restrictive irrigation combined with the application of 250 mgL^{-1} of GA_3 effectively induced flowering in up to 83% of all plants (Cardoso et al. 2010).

Light is another factor that appears to interact with plant hormones. For example, in *Paphiopedilum praestans* Pfitzer plants grown under a light intensity of 60% showed enhanced growth and a treatment of 0.5 mgL^{-1} GA_3 accelerated flowering (Dahlia 2016). Chang et al. (2015) reported that a treatment with GA_3 at 25/15 °C (day/night) quickened the emergence of inflorescences in species and hybrids of *Paphiopedilum* (lady slippers), such as *P.* Ho Chi Minh, *P. liemianum* and *P. primulinum*. Moreover, application of GAs in *Aranda* Deborah can promote development of the vegetative shoot apex into a terminal inflorescence, and in *Bletilla striata*, *Cattleya* and *Cymbidium* GAs also induce flowering (Goh et al. 1982a, b; Goh 1985).

6.8 Regulation of Flowering in Orchids

Based on their growth morphology most orchids can be divided into two main groups: monopodial and sympodial. In orchids that are monopodial, the main vegetative stem emerges from a single bud, elongates and starts to develop leaves close to the apex. Roots emerge continuously from a basal site that anchors the plant (Hew and Yong 2004). Sympodial orchids develop a rhizome at the bottom of the pseudobulb, and grow upward continuously (Hew and Yong 2004). Usually a new pseudobulb may emerge laterally from an old pseudobulb. For monopodial orchids, floral induction in nature may depend on the environmental conditions, in particular cool temperatures or a suitable light intensity. For sympodial orchids, the induction of flowering may depend on cool temperatures, such as in the case of *Dendrobium* of the *nobile* type, or may be induced endogenously when the plant reaches certain maturity (Hew and Yong 2004). Below we discuss the general mechanisms of flowering in the two groups.

1. **Flowering in monopodial orchids**

In *Arabidopsis thaliana*, genes that regulate flowering time have been clearly identified and characterized functionally (Bäurle and Dean 2006; He and Amasino 2005; Jung and Müller 2009). However, in orchids comparatively little work has been done so far, perhaps because of very long life cycles and difficult transformation methodologies. Recently, the identification of potential genes that control orchid flowering has been possible, thanks to genomic studies (Lin et al. 2016; Huang et al. 2016). This type of works benefits from prior work done in *Arabidopsis,* especially in the identification of *EARLY FLOWERING4* (*ELF4*) as a key component of the circadian clock machinery (CCA1/LHY-TOC1) that is directly involved in the transcriptional regulation of the central oscillator and affects flowering through perception of the photoperiod (Doyle et al. 2002; Kikis et al. 2005; Li et al. 2011; McWatters et al. 2007). The ectopic expression of the *Phalaenopsis/Doritaenopsis* homologs of *EFLs* (*DhEFL2*, *DhEFL3* and *DhEFL4*) in *Arabidopsis thaliana* rescues the early-flowering phenotype in the *elf4 homozygous* mutant, while overexpression of *DhEFL2*, *DhEFL3* or *DhEFL4* appears to delay flowering of wild-type *Arabidopsis* plants. Taken together, these results suggest that the *Phalaenopsis/Doritaenopsis* homologs of the *Arabidopsis ELF4* locus might perform similar roles (Chen et al. 2015, 2016).

The *Flowering locus T* (*FT*) is a putative member of the *phosphatidylethanolamine-binding protein* family (PEBP), and it is one of the most relevant integrator genes within the flowering pathway (Kobayashi et al. 1999; Kardailsky et al. 1999; Corbesier et al. 2007). Its orchid homolog named *PaFT1* has been successfully identified and characterized functionally from *Phalaenopsis aphrodite* (Jang et al. 2015). The *PaFT1* homolog is up-regulated during low environmental temperatures that induce flowering, but its expression is not influenced by the photoperiod. Knock-down of *PaFT1* in *Phalaenopsis* through virus-induced gene silencing (VIGS) causes a delay in spiking under low inductive temperatures. It has been shown that PaFT1 physically binds to transcription factor PaFD, which features a basic leucine zipper (bZIP) domain, in a manner similar to the *Arabidopsis* FT-FD complex (Wigge et al. 2005; Abe et al. 2005; Benlloch et al. 2011). Five additional *FT* homologs (*PhFT1* ~ *5*) have been identified in *Phalaenopsis* Tai Lin Red Angel 'V31'. The *PhFT-1* locus shows the highest sequence similarity to the *Arabidopsis FT* across all dicot crops, and when it is expressed in *Arabidopsis* causes early flowering. This phenotype suggests it is a positive regulator of flowering (Zhou et al. 2018).

The LEAFY/FLORICAULA (*LFY/FLO*) is a family of transcription factors that is specific to plants, and is considered to operate both as signal integrators in the floral pathway that regulate the timing of the developmental transition between the vegetative and the reproductive phase. *LFY* itself functions as a determinant of cell fate at the floral meristem (Weigel et al. 1992; Parcy et al. 1998). The putative orchid homolog of *Arabidopsis LFY* is *Phalaenopsis PhalLFY*, which was cloned from the elite *Phalaenopsis* Wedding Promenade hybrid. Analyses by semi-quantitative RT-PCR and by in situ hybridization have revealed that *PhalLFY* accumulates at high levels in the floral stem during the process of floral transition (Zhang et al. 2010). Species *Phalaenopsis aphrodite* subsp. *formosana* has its own *LFY* homolog named *PhapLFY*, which accumulates in the floral primordia of the developing inflorescence. The corresponding PhapLFY protein is a homodimer that can self-activate. Furthermore, ectopic expression of *PhapLFY* in the *Arabidopsis lfy* background can rescue an aberrant floral phenotype. Additionally, transgenic rice expressing *PhapLFY* shows precocious panicle heading (Jang 2015).

The *CONSTANS* (*CO*) and *CONSTANS-like* (*COL*) genes are central regulators of the signaling pathways that sense photoperiod and temperature (Baurle and Dean 2006; Yang et al. 2020). A putative *COL* orthologous gene named *PhalCOL*, was cloned, once again, from *Phalaenopsis* Wedding Promenade. The *PhalCOL* transcript is expressed in all organs during development, and reaches its highest expression level in leaves. During the transition from vegetative to reproductive development, PhalCOL is highly expressed in the floral stem. Ectopic overexpression of the *PhalCOL* transcript in *Nicotiana* causes early flowering, suggesting that there is functional conservation of *CO* gene activity during flowering among plants (Zhang et al. 2011). Ke et al. (2020) identified a *COL1* homolog in *Phalaenopsis aphrodite* subsp. *formosana*. The *PaCOL1* homolog is differentially expressed in leaves, with a peak accumulation during late afternoon and midnight. Overexpression of *PaCOL1* in *Arabidopsis thaliana* causes early flowering under a short-day light regime (8 h of light). Results from yeast two-hybrid experiments and by bimolecular fluorescence complementation (BiFC) indicate that the PaCOL1 protein is able to interact with the circadian clock regulator, AtCCA1, and with the suppressor of flowering, AtFLC (Ke et al. 2020).

The *MULTICOPY SUPPRESSOR OF IRA 4* (*FVE/MSI4*) is a gene that shows high evolutionary conservation, and in *Arabidopsis thaliana* it operates as a regulator of flowering time within the autonomous flowering pathway (Hennig et al. 2005; Ausín et al. 2004). A putative *FVE* homolog was identified and cloned from *Doritaenopsis* 'Tinny Tender'. *DhFVE* shows high transcriptional activity in roots, leaves and stems during the developmental transition from vegetative growth to flowering, with the highest activity level in flower stem.

Expression of *DhFVE* in *Arabidopsis thaliana* causes late flowering under both long and short-day conditions, a function similar to that of *FVE* (Sun et al. 2012). Moreover, Koh et al. (2018) were able to identify an *FVE* homolog from the species *Phalaenopsis aphrodite* subsp. *formosana*. The *PaFVE* transcript is inducible by low environment temperatures, and high expression levels are observed after initial stage of spiking initiation and remains very high during the early stages of flower development. Endogenous knockdown of *PaFVE* expression using the VIGS system caused a delay in the maturation of flower buds but did not disrupt the floral transition. The delay in the maturation of flower buds is linked to cessation of organ growth, suggesting that possibly *PaFVE* might intervene in the early stages of developmental regulation of flower growth. In transgenic *Arabidopsis* the expression of *PaFVE* led to quick flowering, and a concomitant increase in the transcription of *AtSOC1*, suggesting that *PaFVE* is a *bona fide* regulator of floral development (Koh et al. 2018). Double floral spikes in *Phalaenopsis* potted plants are ideal for the market. In *Phalaenopsis aphrodite* subsp. *formosana*, Lin et al. (2019) identified the *Spike Activator 1* (*SPK1*) gene, which appears to encode a transcription factor featuring a basic helix-loop-helix (bHLH) motif. The expression of *PaSPK1* increases significantly when the temperature reaches 20 °C (the temperature for spiking induction) but it is suppressed at 30 °C, a temperature unsuitable for spiking. Coincidentally, the *PaSPK1* transcript was found to show a high expression pattern in floral meristems. Knockdown of endogenous *PaSPK1* expression using a VIGS system showed that in VIGS-*spk1* plants production of double spikes occurs at a low rate of 1/15, suggesting a positive role for *PaSPK1* during flower spike initiation (Lin et al. 2019). In order to broaden the list of candidate genes that may regulate orchid flowering, the species *Phalaenopsis aphrodite* had its axillary buds analyzed to uncover transcriptional patterns. Those axillary buds were collected from plants exposed to low and high temperatures, and the results were contrasted with each other. The results show that the putative homologs of *LFY*, *FT*, *APETALA1* (*AP1*), *SUPPRESSOR OF OVER-EXPRESSION OF CONSTANS 1* (*SOC1*) and several genes required for the synthesis of GAs are robustly up-regulated by low temperatures that are well-known to promote spiking (Huang et al. 2016) (Fig. 6.3).

Vanda is a very important genus of tropical monopodial orchids mostly cultivated in Thailand as potted plants and for the cut flower market and with a great potential in international markets. The putative *Vanda* homolog of *Arabidopsis FT*, aptly named *VaFT*, was cloned from *Vanda* 'Ratchaburi-Fusch Katsura'. Overexpression of *VaFT* in transgenic *Arabidopsis* plants led to transcriptional activation of *APETALA1* (*AP1*), a determinant of cell fate in the floral meristem, and also caused early flowering. This finding implies that *VaFT* is a putative PEBP factor and may regulate the transition to flowering in *Vanda* (Panjama et al. 2019).

Plant miRNAs play critical roles in several developmental processes, such as hormonal regulation, organ polarity, organ boundary formation, nutrient homeostasis, stress responses, and development of vegetative and reproductive organs (Rhoades et al. 2002; Kidner and Martienssen 2005; Jones-Rhoades et al. 2006; Husbands et al. 2009). Transcriptional analysis by high-throughput sequencing of sRNA was performed in *Phalaenopsis aphrodite* in order to determine how miRNAs respond to low environmental temperatures. Results from the sequencing experiments and from Northern hybridization of smRNAs showed that transcripts *miR156*, *miR162*, *miR528* and *miR535* are induced by low temperature. Moreover, the spatiotemporal specificity of expression in these miRNAs suggests that in *Phalaenopsis miR156* and *miR172* might be key components of a regulatory pathway that mediates the transition from the vegetative to the reproductive phase (An et al. 2011). Similar control of reproductive phase change by miR156 and miR172 also operates in tropical and subtropical tree crops (Ahsan et al. 2019).

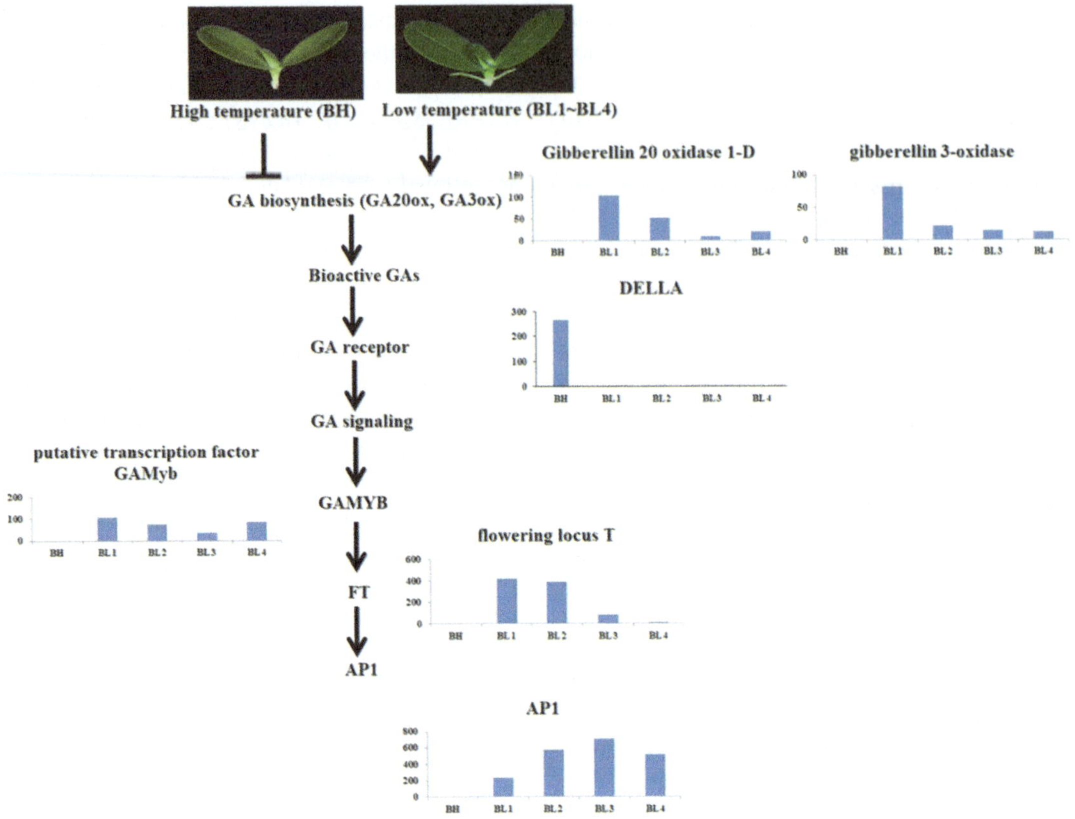

Fig. 6.3 Involvement of a probable GA response pathway in *Phalaenopsis aphrodite* may contribute to induce flower spikes, as suggested by analysis of the transcriptome (modified after Huang et al. 2016)

2. **Sympodial orchids**

The *nobile*-type *Dendrobiums* are sympodial orchids and are one of the most commercially traded orchids globally because of their ornamental and medicinal value. These are biennial plants and require a vernalization period for flowering (Campos and Kerbauy 2004). In *Dendrobium nobile* it has been possible to identify a putative homolog of *Arabidopsis FT* named *DnFT* and *MOTHER OF FT AND TFL1* (*MFT*) (Li et al. 2012). Under low temperature (12 °C/9 °C, day/night) *DnFT* is up-regulated in leaves but down-regulated in axillary buds, while the expression of *DnMFT* is not influenced by low temperature. A temperature regime of 12 °C/ 9 °C, day/night causes *DnFT* to be highly expressed in young buds and comparatively less in mature buds; however, down-regulation is observed in axillary buds. During flower bud development the expression of *DnMFT* was observed to increase in floral buds and decrease in leaves. In *Arabidopsis* the ectopic expression of *DnFT* causes early flowering. Overexpression of *DnMFT* in *Arabidopsis* causes a slight late-flowering phenotype (Li et al. 2012).

In the model plant *Arabidopsis*, the *SOC1* gene encodes a MADS-box protein that plays an indispensable role during the integration of multiple signals that elicit the transition from vegetative to reproductive development, during and floral patterning and floral meristem determinacy (Lee and Lee 2010). *DOSOC1* is a putative *SOC1-like* gene, identified and isolated from hybrid *Dendrobium* Chao Praya Smile (Ding et al. 2013). *DOSOC1* shows a pattern of high expression in reproductive organs, including the inflorescence apices, the pedicels, the floral buds and flowers after anthesis. Its expression increases significantly during the floral transition in whole plantlets that are grown in vitro. When overexpressed in the *Arabidopsis soc1-2* background, *DOSOC1* was shown to partially rescue a late-flowering phenotype. Moreover, overexpression of *DOSOC1* in both *Arabidopsis thaliana* and *Dendrobium* Chao Praya Smile results in early-flowering phenotypes, suggesting that the activity *DOSOC1* is roughly equivalent to that of *AtSOC1* in terms of promoting flowering (Ding et al. 2013).

In *Arabidopsis* the *AP1* gene encodes another MADS-box transcription factor that specifies cell identity in the floral meristem and determines the identity of cells in the perianth (Irish and Sussex 1990; Alejandra-Mandel et al. 1992; Bowman et al. 1993). Ectopic expression of *DOAP1*, an ortholog of *AP1* identified in *Dendrobium* Chao Praya Smile, is able to rescue developmental defects in flowers of the *Arabidopsis ap1* mutant but results in early flowering of wild-type *Arabidopsis* plants. Overexpression of *DOAP1* in *Dendrobium* led to early flowering and precocious cessation of the transition of the inflorescence meristem into a floral meristem, when compared to the wild-type plants. Thus, *DOAP1* has been proposed to have retained its role as a promoter of flowering and a key determinant of cell identity and fate at the floral meristem in *Dendrobium* (Sawettalake et al. 2017).

DOH1 and *DOMADS1* are genes identified in *Dendrobium* Madame Thong-In. The *DOH1* gene seems to encode a class-1 *knox* factor that might play a repressive role during the formation and development of the shoot apical meristem (SAM) and during the determination of shoot architecture, while *DOMADS1* expression may behave as a marker of the floral transition in the SAM (Yu et al. 2000).

Overexpression of an antisense *DOH1* transcript in *Dendrobium* causes early flowering, suggesting that it behaves as a repressor of *DOMADS1* activity during the regulation of flowering (Yu and Goh 2000; Yu et al. 2000). A transcriptome study by RNA-seq analysis in *Dendrobium nobile* identified several differentially expressed genes (DEGs) when plants are exposed to thidiazuron (TDZ, a cytokinin) or cold (4 °C) (Wen et al. 2017). Several genes involved in the transition to flowering such as *DnVRN1*, *DnFT*, *DnSOC1*, *DnLFY* and *DnAP1* were differentially expressed in these plants, compared to the mock. Both cold and TDZ boosted the expression of *DnSOC1*, *DnLFY* and *DnAP1*, while the up-regulation of *DnVRN1* and *DnFT* was only caused by cold. Also, several key genes for the GA signaling pathway were up-regulated under both treatments. This study suggests that there is a level of cross-talk between the cytokinin and GA signaling networks during vernalization, suggesting the existence of a novel molecular mechanism that mediates the initiation of flowering in *Dendrobium* (Wen et al. 2017).

In *Oncidium* (now *Oncidesa*) Gower Ramsey, it has been possible to identify several genes that might be involved in the regulation of flowering time. For instance, the *OMADS1* gene is believed to be *AP1/AGL9* MADS-box gene, which shows high homology to the famous *Arabidopsis AGAMOUS-like 6* gene (*AGL6*) (Hsu et al. 2003). *OMADS1* significantly promotes flowering by facilitating the ectopic activation of *FT*, *SOC1 LFY* and *AP1* in transgenic *Arabidopsis* plants (Hsu et al. 2003). Another gene that was isolated and characterized from *Oncidium* is *OMADS3,* which may be responsible for the regulation of flower formation and floral initiation (Hsu and Yang 2002). Later on, four additional *AP1/AGL9-like* genes from *Oncidium* enlarged the family; they are *OMADS6*, *OMADS7*, *OMADS10* and *OMADS11* (Chang et al. 2009). Ectopic expression in *Arabidopsis* of *OMADS6*, *7* and *11* causes extremely early flowering, whereas expression of *OMADS10* causes a slight early-flowering phenotype (Chang et al. 2009). Suspected orthologs of *Arabidopsis FT* have been isolated and characterized from *Oncidium* Gower Ramsey, and they are called *OnFT* and *OnTFI* (*TERMINAL FLOWER 1*). The *OnFT* transcript can be detected in axillary buds, leaves, pseudobulbs and even flowers. The expression of *OnFT* is regulated by the photoperiod, and it is at its lowest during daybreak and peaks after 8–12 h of light. However, expression of *OnTFL1* is not influenced by the photoperiod and its expression is only detectable in axillary buds and pseudobulbs. Ectopic expression of *OnFT* causes early flowering in *Arabidopsis thaliana*, and rescues the late-flowering phenotype of the *ft-1* mutant. Ectopic expression of *OnTFL1* in the transgenic *Arabidopsis tfl1-11* mutant also delays flowering and rescues a terminal flower phenotype (Hou and Yang 2009).

Research in the spring orchid (*Cymbidium goeringii*) allowed identification of a putative *CgFT* transcript in leaves, pseudobulbs and flowers. The *CgFT* transcript shows a peak of expression in young flower buds, and very high expression in young ovaries (Xiang et al. 2012). Taken together, the available evidence shows that *FT* orthologs in orchids appear to show distinct spatiotemporal expression patterns and differ in their regulatory mechanisms. However, ectopic expression of these orchid *FT* orthologs in transgenic *Arabidopsis* and *Nicotiana* plants caused the same extremely early-flowering phenotype (Hou and Yang 2009; Xiang et al. 2012; Jang 2015), indicating that they all share a universal function across different plant species.

Several putative regulators of flowering have been identified in the genus *Cymbidium*, thanks to analysis of the transcriptome, such as in the case of species *Cymbidium sinense* (Zhang et al. 2013), *Cymbidium faberi* Rolfe (Sun et al. 2016) and *Cymbidium goeringii* (Yang et al. 2019). Three homologs of the *Arabidopsis thaliana SHORT VEGETATIVE PHASE* (*SVP*) gene were identified in *Cymbidium goeringii* and appear to belong to different evolutionary sub-clades (Yang et al. 2019). The *Arabidopsis SVP* gene encodes a MADS-box transcription factor that functions as a repressor of the expression of other genes that integrate flowering signals, and as a result the floral transition is delayed by *SVP* (Hartmann et al. 2000). A cold treatment may inhibit the expression of the three *CgSVP* homologs. Besides, ectopic expression of these putative *CgSVP* genes in other *Cymbidium* varieties with different flowering behavior leads to uniform flowering after exposure to cold. These *CgSVP* genes show transcriptionally up-regulation by cold earlier than *CgSOC1*, *CgLFY* and *CgAP1*, suggesting the existence of a response to vernalization that is reminiscent to that shown by the *Arabidopsis FLC* gene. Thus, *CgSVPs* seem to be important regulators upstream of other genes (Yang et al. 2019). A CgSVP protein was shown to interact physically with CgAP1 and CgSOC1 by Y2H analysis. Therefore, these genes may regulate flowering synergistically in *Cymbidium goeringii* during the winter (Yang et al. 2019).

6.9 Conclusions and Perspectives

There has been noticeable progress in the understanding of flowering in orchids, thanks to detailed mechanistic research into flower induction in *Phalaenopsis*, *Cymbidium* and *Dendrobium*. However, only limited information is available about the regulation of processes that control flower induction and organ development in most other orchids. Extensive research is still needed to develop commercially sound methods for the control of flowering in economically relevant orchid genera such as *Paphiopedilum*, *Oncidium*, *Aranda*, *Vanda* and *Dendrobium* with the aim of assisting current orchid breeding programs. Whole-genome sequencing has been performed in *Phalaenopsis equestris* (Cai et al. 2015), the cultivar *Phalaenopsis* 'KHM190' (Huang et al. 2016), *Dendrobium catenatum* (Zhang et al. 2016), *Dendrobium officinale* (Yan et al. 2015), and *Apostasia shenzhenica* (Zhang et al. 2017), and it is hoped that the genetic knowledge gained may contribute to efforts to control flowering.

References

Abe M, Kobayashi Y, Yamamoto S, Daimon Y, Yamaguchi A, Ikeda Y, Ichinoki H, Notaguchi M, Goto K, Araki T (2005) FD, a bZIP protein mediating signals from the floral pathway integrator FT at the shoot apex. Science 309:1052–1056

Abraham A, Vatsala P (1981) Introduction to Orchids with Illustrations and Descriptions of 150 South Indian Orchids. TBGRI, Trivandrum, Kerala, India

Ahn JH, Miller D, Winter VJ, Banfield MJ, Lee JH, Yoo SY, Henz SR, Brady RL, Weigel D (2006) A divergent external loop confers antagonistic activity on floral regulators FT and TFL1. EMBO J 25:605–614

Ahsan MU, Hayward A, Irihimovitch V, Fletcher S, Tanurdzic M, Pocock A, Beveridge CA, Mitter N (2019) Juvenility and vegetative phase transition in tropical/subtropical tree crops. Front Plant Sci 10:729

Alejandra Mandel M, Gustafson-Brown C, Savidge B, Yanofsky MF (1992) Molecular characterization of the *Arabidopsis* floral homeotic gene APETALA1. Nature 360:273–277

An FM, Hsiao SR, Chan MT (2011) Sequencing-based approaches reveal low ambient temperature-responsive and tissue-specific microRNAs in *Phalaenopsis* orchid. PLoS One 6:e18937

Andrés F, Coupland G (2012) The genetic basis of flowering responses to seasonal cues. Nat Rev Genet 13:627–639

Arditti J (1979) Aspects of the Physiology of Orchids. Academic Press, London

Ashikari M, Sakakibara H, Lin S, Yamamoto T, Takashi T, Nishimura A, Angeles ER, Qian Q, Kitano H, Matsuoka M (2005) Cytokinin oxidase regulates rice grain production. Science 309:741–745

Audus LJ (1972) Plant Growth Substances I. Chemistry and Physiology, Leonard Hill, London

Ausín I, Alonso-Blanco C, Jarillo JA, Ruiz-García L, Martínez-Zapater JM (2004) Regulation of flowering time by FVE, a retinoblastoma-associated protein. Nat Genet 36:162–166

Bünning E (1936) Die endogene tagesrhythmik als grundlage der photoperiodischen reaktion. Ber Dtsch Bot Ges 54:590–607

Baker ML, Baker CO (1993) *Miltoniopsis* culture II. Amer Orchid Soc Bull 62:901–908

Baker ML, Baker CO (1996) Orchid Species Culture: *Dendrobium*. Timber Press, Portland

Bartrina I, Otto E, Strnad M, Werner T, Schmülling T (2011) Cytokinin regulates the activity of reproductive meristems, flower organ size, ovule formation, and thus seed yield in *Arabidopsis thaliana*. Plant Cell 23:69–80

Batchelor RS (2011) Beginner's series, Orchid culture: Part 4, Light and temperature. https://staugorchid-society.org/PDF/AOS4-LightTemp.pdf

Baurle I, Dean C (2006) The timing of developmental transitions in plants. Cell 125:655–664

Benlloch R, Kim MC, Sayou C, Thévenon E, Parcy F, Nilsson O (2011) Integrating long-day flowering signals: a LEAFY binding site is essential for proper photoperiodic activation of APETALA1. Plant J 67:1094–1102

Bernier G (1988) The control of floral evocation and morphogenesis. Annu Rev Plant Physiol Plant Mol Biol 39:175–219

Bernier G, Havelange A, Houssa C, Petitjean A, Lejeune P (1993) Physiological signals that induce flowering. Plant Cell 5:1147–1155

Bernier G, Kinet J, Sachs RM (1981) The Physiology of Flowering, vol. 1. CRC Press, Boca Raton, pp 106–107

Blanchard MG, Runkle ES (2006) Temperature during the day, but not during the night, controls flowering of *Phalaenopsis* orchids. J Exp Bot 57:4043–4049

Blanchard MG, Runkle ES (2005) Temperature regulates flowering of two *Odontioda* orchid hybrids. HortScience 40:1404–1409

Blanchard MG, Runkle ES (2008a) Benzyladenine promotes flowering in *Doritaenopsis* and *Phalaenopsis* orchids. J Plant Growth Regul 27:141

Blanchard MG, Runkle ES (2008b) Temperature and pseudobulb size influence flowering of *Odontioda* orchids. HortScience 43:1404–1409

Bouché F, Lobet G, Tocquin P, Périlleux C (2016) FLOR-ID: an interactive database of flowering-time gene networks in *Arabidopsis thaliana*. Nucleic Acids Res 44:D1167–D1171

Bowman JL, Alvarez J, Weigel D, Meyerowitz EM, Smyth DR (1993) Control of flower development in *Arabidopsis thaliana* by APETALA1 and interacting genes. Development 119:721–743

Cai J, Liu X, Vanneste K, Proost S, Tsai WC, Liu KW, Chen LJ, He Y, Xu Q, Bian C, Zheng Z, Sun F, Liu W, Hsiao YY, Pan ZJ, Hsu CC, Yang YP, Hsu YC, Chuang YC, Dievart A, Dufayard JF, Xu X, Wang JY, Wang J, Xiao XJ, Zhao XM, Du R, Zhang GQ, Wang M, Su YY, Xie GC, Liu GH, Li LQ, Huang LQ, Luo YB, Chen HH, Van de Peer Y, Liu ZJ (2015) The genome sequence of the orchid *Phalaenopsis equestris*. Nat Genet 47:65–72

Campos KO, Kerbauy GB (2004) Thermoperiodic effect on flowering and endogenous hormonal status in *Dendrobium* (Orchidaceae). J Plant Physiol 161:1385–1387

Cardoso JC, Ono EO, Rodrigues JD (2010) Gibberellic acid and water regime in the flowering induction of *Brassocattleya* and *Cattleya* hybrid orchids. Hortic Bras 28:395–398

Chandler J, Dean C (1994) Factors influencing the vernalization response and flowering time of late flowering mutants of *Arabidopsis thaliana* (L.) Heynh. J Exp Bot 45:1279–1288

Chang CC, Yang AH, Wang SS (2015) Studies on the regulation of flowering of *Paphiopedilum* orchids. Res Bull Tainan Dist Agri Improv Stn 65:10–19 (in Chinese)

Chang YY, Chiu YF, Wu JW, Yang CH (2009) Four orchid (Oncidium Gower Ramsey) AP1/AGL9-like MADS box genes show novel expression patterns and cause different effects on floral transition and formation in *Arabidopsis thaliana*. Plant Cell Physiol 50:1425–1438

Chen WS, Chang HW, Chen WH, Lin YS (1997) Gibberellic acid and cytokinin affect *Phalaenopsis* flower morphology at high temperature. HortScience 32:1069–1073

Chen WS, Liu HY, Liu ZH, Yang L, Chen WH (1994) Geibberllin and temperature influence carbohydrate content and flowering in *Phalaenopsis*. Physiol Plant 90:391–395

Chen W, Qin Q, Zhang C, Zheng Y, Wang C, Zhou M, Cui Y (2015) DhEFL2, 3 and 4, the three EARLY FLOWERING4-like genes in a *Doritaenopsis* hybrid regulate floral transition. Plant Cell Rep 34:2027–2041

Chen W, Qin Q, Zheng Y, Wang C, Wang S, Zhou M, Zhang C, Cui Y (2016) Overexpression of *Doritaenopsis* hybrid EARLY FLOWERING 4-like4 gene, DhEFL4, postpones flowering in transgenic *Arabidopsis*. Plant Mol Biol Rep 34:103–117

Cho LH, Yoon J, An G (2017) The control of flowering time by environmental factors. Plant J 90:708–719

Corbesier L, Coupland G (2005) Photoperiodic flowering of *Arabidopsis*: integrating genetic and physiological approaches to characterization of the floral stimulus. Plant Cell Environ 28:54–66

Corbesier L, Vincent C, Jang S, Fornara F, Fan Q, Searle I, Giakountis A, Farrona S, Gissot L, Turnbull C, Coupland G (2007) FT protein movement contributes to long-distance signaling in floral induction of *Arabidopsis*. Science 316:1030–1033

Craufurd PQ, Wheeler TR (2009) Climate change and the flowering time of annual crops. J Exp Bot 60:2529–2539

Cumming BG (1959) Extreme sensitivity of germination and photoperiodic reaction in the genus *Chenopodium* (Tourn.) L. Nature 184:1044–1045

Dahlia (2016) Paphiopedilum flowering induction with light intensity and growth regulator substance. In: AIP Conference Proceedings 1744: 020010

Dalström S, Higgins WE (2016) New combinations and transfers to *Odontoglossum* Oncidiinae (Orchidaceae): avoid creating new names. Harvard Papers in Botany 21:97–104

Damann MP, Lyons RE (1993) Juvenility, flowering and the effects of a limited inductive photoperiod in *Coreopsis grandiflora* and *C. landceolata*. J Am Soc Hortic Sci 118:513–518

Davière JM, Achard P (2013) Gibberellin signaling in plants. Development 140:1147–1151

Day J, Loveys B, Aspinall D (1995) Cytokinin and carbohydrate changes during flowering of *Boronia megastigma*. Funct Plant Biol 22:57–65

de Melo Ferreira W, Barbante Kerbauy G, Elizabeth Kraus J, Pescador R, Mamoru Suzuki R (2006) Thidiazuron influences the endogenous levels of cytokinins and IAA during the flowering of isolated shoots of *Dendrobium*. J Plant Physiol 163:1126–1134

De Vries JT (1953) On the flowering of *Phalaenopsis schilleriana* RCHB. f. Annu Bogor 1:61–76

Dewitte W, Chiappetta A, Azmi A, Witters E, Strnad M, Rembur J, Noin M, Chriqui D, Van Onckelen H (1999) Dynamics of cytokinins in apical shoot meristems of a day-neutral tobacco during floral transition and flower formation. Plant Physiol 119:111–122

Ding L, Wang Y, Yu H (2013) Overexpression of DOSOC1, an ortholog of *Arabidopsis* SOC1, promotes flowering in the orchid *Dendrobium* Chao Parya Smile. Plant Cell Physiol 54:595–608

Distelfeld A, Li C, Dubcovsky J (2009) Regulation of flowering in temperate cereals. Curr Opin Plant Biol 12:178–184

Doyle MR, Davis SJ, Bastow RM, McWatters HG, Kozma-Bognár L, Nagy F, Millar AJ, Amasino RM (2002) The ELF4 gene controls circadian rhythms and flowering time in *Arabidopsis thaliana*. Nature 419:74–77

Dressler RL, Higgins WE (2003) *Guarianthe*, a generic name for the "*Cattleya*" *skinneri* complex. Lankesteriana 7:37–38

Eigeldinger O, Murphy LS (1972) Orchids: a Complete Guide to Cultivation. Drake Publ, New York

Friend DJC (1968) Photoperiodic responses of *Brassica campestris* cv Ceres. Physiol Plant 21:990–1002

Garner WW, Allard HA (1920) Effect of the relative length of day and night and other factors of the environment of growth and reproduction in plants. J Agric Res 18:553–603

Gao J, Liang D, Xu Q, Yang F, Zhu G (2020) Involvement of *CsERF2* in leaf variegation of *Cymbidium sinense* 'Dharma'. Planta 252:29

Goh CJ (1977) Regulation of floral initiation and development in an orchid hybrid *Aranda* Deborah. Ann Bot 41:763–769

Goh CJ (1979) Hormonal regulation of flowering in a sympodial orchid hybrid *Dendrobium* Louisae. New Phytol 82:375–380

Goh CJ (1985) Flowering in tropical orchids. In: Proceedings of XIth World Orchid Conf 17 May 1982, Florida, pp 166–173

Goh CJ, Wan HY (1973) The role of auxins in flowering of a tropical orchid hybrid *Vanda* X Miss. Sumiki Y (ed) Plant Growth Substances, Hirotaka Publ Co, Tokyo 91:945–952

Goh CJ, Yang AL (1978) Effects of growth regulators and decapitation on flowering of *Dendrobium* orchid hybrids. Plant Sci Lett 12:278–292

Goh CJ, Strauss MS, Arditti J, Arditti J (1982a) Flower Induction and Physiology in Orchids. Cornell University Press, Ithaca

Goh CJ, Strauss MS, Arditti J (1982b) Flower induction and physiology in orchids. In: Arditti J (ed) Orchid Biology: Reviews and Perspectives, II. Cornell University Press, Ithaca, NY, pp 213–241

Gray WM (2004) Hormonal regulation of plant growth and development. PLoS Biol 2:e311

Griesbach RJ (1985) An orchid in every pot. Florists' Rev 176:26–30

Hackett WP (1986) Juvenility, maturation, and rejuvenation in woody plants. Hortic Rev 7:109–155

Hartmann U, Höhmann S, Nettesheim K, Wisman E, Saedler H, Huijser P (2000) Molecular cloning of SVP: a negative regulator of the floral transition in *Arabidopsis*. Plant J 21:351–360

Hayama R, Coupland G (2004) The molecular basis of diversity in the photoperiodic flowering responses of *Arabidopsis* and rice. Plant Physiol 135:677–684

He Y, Amasino RM (2005) Role of chromatin modification in flowering-time control. Trends Plant Sci 10:30–35

Hedden P, Sponsel V (2015) A century of gibberellin research. J Plant Growth Regul 34:740–760

Hennig L, Bouveret R, Gruissem W (2005) MSI1-like proteins: an escort service for chromatin assembly and remodeling complexes. Trends Cell Biol 15:295–302

Hew CS, Clifford PE (1993) Plant growth regulators and the orchid cut-flower industry. Plant Growth Regul 13:231–239

Hew CS, Yong JWH (2004) The Physiology of Tropical Orchids in Relation to the Industry, 2nd edn. World Scientific, Singapore

Higuchi H, Sakai K (1977) The effect of N^6-benzyladenine on the flowering of *Dendrobium* Nodoka. Res Bull Aichi-Ken Agric Res Cent, Series B 9:79–81

Holttum RE (1949) Gregarious flowering of the terrestrial orchid *Bromheadia finlaysoniana*. Gard Bull (Singapore) 12:295–302

Hou CJ, Yang CH (2009) Functional analysis of FT and TFL1 orthologs from orchid (*Oncidium* Gower Ramsey) that regulate the vegetative to reproductive transition. Plant Cell Physiol 50:1544–1557

Hsu HF, Huang CH, Chou LT, Yang CH (2003) Ectopic expression of an orchid (*Oncidium* Gower Ramsey) AGL6-like gene promotes flowering by activating flowering time genes in *Arabidopsis thaliana*. Plant Cell Physiol 44:783–794

Hsu HF, Yang CH (2002) An orchid (*Oncidium* Gower Ramsey) AP3-like MADS gene regulates floral formation and initiation. Plant Cell Physiol 43:1198–1209

Huang JZ, Lin CP, Cheng TC, Huang YW, Tsai YJ, Cheng SY, Chen YW, Lee CP, Chung WC, Chang BCH, Chin SW, Lee CY, Chen FC (2016) The genome and transcriptome of *Phalaenopsis* yield insights into floral organ development and flowering regulation. PeerJ 4:e2017

Husbands AY, Chitwood DH, Plavskin Y, Timmermans MCP (2009) Signals and prepatterns: new insights into organ polarity in plants. Genes Dev 23:1986–1997

Ichihashi S, Arditti J, Pridgeon AM (1997) Orchid Production and Research in Japan. Kluwer Academic Publisher

Imaizumi T, Kay SA (2006) Photoperiodic control of flowering: not only by coincidence. Trends Plant Sci 11:550–558

Irish VF, Sussex IM (1990) Function of the apetala-1 gene during *Arabidopsis* floral development. Plant Cell 2:741–753

Jacqmard A, Detry N, Dewitte W, Van Onckelen H, Bernier G (2002) In situ localisation of cytokinins in the shoot apical meristem of *Sinapis alba* at floral transition. Planta 214:970–973

Jaeger KE, Graf A, Wigge PA (2006) The control of flowering in time and space. J Exp Bot 57:3415–3418

Jagadish SVK, Bahuguna RN, Djanaguiraman M, Gamuyao R, Prasad PVV, Craufurd PQ (2016) Implications of high temperature and elevated CO_2 on flowering time in plants. Front Plant Sci 7:913

Jang S (2015) Functional characterization of PhapLEAFY, a FLORICAULA/LEAFY ortholog in *Phalaenopsis aphrodite*. Plant Cell Physiol 56:2234–2247

Jang S, Choi SC, Li HY, An G, Schmelzer E (2015) Functional characterization of *Phalaenopsis aphrodite* flowering genes PaFT1 and PaFD. PLoS One 10: e0134987

Johansson M, Staiger D (2014) Time to flower: interplay between photoperiod and the circadian clock. J Exp Bot 66:719–730

Jones-Rhoades MW, Bartel DP, Bartel B (2006) MicroRNAs and their regulatory roles in plants. Annu Rev Plant Biol 57:19–53

Jung C, Müller AE (2009) Flowering time control and applications in plant breeding. Trends Plant Sci 14:563–573

Kardailsky I, Shukla VK, Ahn JH, Dagenais N, Christensen SK, Nguyen JT, Chory J, Harrison MJ, Weigel D (1999) Activation tagging of the floral inducer FT. Science 286:1962–1965

Ke YT, Lin KF, Gu CH, Yeh CH (2020) Molecular characterization and expression profile of PaCOL1, a CONSTANS-like gene in *Phalaenopsis* orchid. Plants (Basel) 9:E68

Kieber JJ, Schaller GE (2018) Cytokinin signaling in plant development. Development 145:dev149344

Kidner CA, Martienssen RA (2005) The developmental role of microRNA in plants. Curr Opin Plant Biol 8:38–44

Kieffer M, Davies B (2001) Developmental programmes in floral organ formation. Semin Cell Dev Biol 12:373–380

Kikis EA, Khanna R, Quail PH (2005) ELF4 is a phytochrome-regulated component of a negative-feedback loop involving the central oscillator components CCA1 and LHY. Plant J 44:300–313

Kim TJ, Lee CH, Paek KY (2000) Effects of growth regulators under low temperature environment on growth and flowering of *Doritaenopsis* 'Happy Valentine' during summer. J Kor Soc Hortic Sci 41:101–104

Kim YJ, Lee HJ, Kim KS (2011) Night interruption promotes vegetative growth and flowering of *Cymbidium*. Sci Hortic 130:887–893

Klebs G (1918) Über die blütenbildung von sempervivum. Flora oder Allgemeine Bot Zeit 111–112:128–151

Kobayashi Y, Kaya H, Goto K, Iwabuchi M, Araki T (1999) A pair of related genes with antagonistic roles in mediating flowering signals. Science 286:1960–1962

Koehn A (1952) Chinese flower symbolism. Monumenta Nipponica 8:121–146. https://www.jstor.org/stable/2383008

Koh KW, Lee S-H, Chen HK, Chang CY, Chan MT (2018) *Phalaenopsis* flowering locus VE regulates floral organ maturation. Plant Cell Rep 37:467–482

Kostenyuk I, Oh BJ, So IS (1999) Induction of early flowering in *Cymbidium niveo-marginatum* Mak *in vitro*. Plant Cell Rep 19:1–5

Kyozuka J (2007) Control of shoot and root meristem function by cytokinin. Curr Opin Plant Biol 10:442–446

Langridge J (1957) Effect of day-length and gibberellic acid on the flowering of *Arabidopsis*. Nature 180:36–37

Lee N, Lin GM, Chang LR (1987) Controlling the flowering of Phalaenopsis. In: Paper presented at the proceedings on symposium forcing Cult Hortic Crops Special Publ 10. Taichung District Agr. Improvement Stn (in Chinese)

Lee J, Lee I (2010) Regulation and function of SOC1, a flowering pathway integrator. J Exp Bot 61:2247–2254

Lee J, Oh M, Park H, Lee I (2008) SOC1 translocated to the nucleus by interaction with AGL24 directly regulates LEAFY. Plant J 55:832–843

Lee YR, Lee DW, Won JY, Kim MS, Kim JY, Lee JS (1998) Effect of BA on flowering of *Cymbidium ensifolium* 'Tekkotsusosin'. Korean J Hortic Sci Technol 16:531–532

Leonhardt KW (2000) Potted blooming Dendrobium orchids. HortTechnology 10:431–432

Li G, Siddiqui H, Teng Y, Lin R, Wan XY, Li J, Lau OS, Ouyang X, Dai M, Wan J, Devlin PF, Deng XW, Wang H (2011) Coordinated transcriptional regulation underlying the circadian clock in *Arabidopsis*. Nat Cell Biol 13:616–622

Li R, Wang A, Sun S, Liang S, Wang X, Ye Q, Li H (2012) Functional characterization of FT and MFT ortholog genes in orchid (*Dendrobium nobile* Lindl) that regulate the vegetative to reproductive transition in *Arabidopsis*. Plant Cell Tissue Organ Cult 111:143–151

Lin CS, Hsu CT, Liao DC, Chang WJ, Chou ML, Huang YT, Chen JJW, Ko SS, Chan MT, Shih MC (2016) Transcriptome-wide analysis of the MADS-box gene family in the orchid *Erycina pusilla*. Plant Biotechnol J 14:284–298

Lin YJ, Li MJ, Hsing HC, Chen TK, Yang TT, Ko SS (2019) Spike Activator 1, encoding a bHLH, mediates axillary bud development and spike initiation in *Phalaenopsis aphrodite*. Int J Mol Sci 20:E5406

Lohmann JU, Hong RL, Hobe M, Busch MA, Parcy F, Simon R, Weigel D (2001) A molecular link between stem cell regulation and floral patterning in *Arabidopsis*. Cell 105:793–803

Lopez RG, Runkle ES (2004a) The effect of temperature on leaf and flower development and flower longevity of *Zygopetalum redvale* 'Fire Kiss' orchid. HortScience 39:1630–1634

Lopez RG, Runkle ES (2004b) The flowering of orchids. Orchids (March 2004) pp 196–203

Lopez RG, Runkle ES (2004) The effect of temperature on leaf and flower development and flower longevity of *Zygopetalum* Redvale 'Fire Kiss' orchid. HortScience 39:1630–1634

Lopez RG, Runkle ES (2005) Environmental physiology of growth and flowering of orchids. HortScience 40:1969–1973

Lopez RG, Runkle ES (2006) Temperature and photoperiod regulate flowering of potted *Miltoniopsis* orchids. HortScience 41:593–597

Lopez RG, Runkle ES, Heins RD, Whitman CM (2003) Temperature and photoperiodic effects on growth and flowering of *Zygopetalum redvale* 'Fire Kiss' orchids. Acta Hortic 624:155–162

Lumsden PJ (2002) Photoperiodism in plants. In: Kumar V (ed) Biological Rhythms. Springer, Berlin Heidelberg, pp 181–191

Macháčková I, Krekule J, Eder J, Seidlová F, Strnad M (1993) Cytokinins in photoperiodic induction of flowering in *Chenopodium* species. Physiol Plant 87:160–166

Matsumoto TK (2006) Gibberellic acid and benzyladenine promote early flowering and vegetative growth of *Miltoniopsis* orchid hybrids. HortScience 41:131–135

McWatters HG, Kolmos E, Hall A, Doyle MR, Amasino RM, Gyula P, Nagy F, Millar AJ, Davis SJ (2007) ELF4 is required for oscillatory properties of the circadian clock. Plant Physiol 144:391–401

Mercier H, Endres L (1999) Alteration of hormonal levels in a rootless epiphytic bromeliad in different phenological phases. J Plant Growth Regul 18:121–125

Mok DW, Mok MC (2001) Cytokinin metabolism and action. Annu Rev Plant Physiol Plant Mol Biol 52:89–118

Mor Y, Halevy AH (1979) Translocation of ^{14}C-assimilates in roses. Physiol Plant 45:177–182

Morrison A (2000) The Illustrated Encyclopedia of Orchids. Timber Press, Portland

Mouradov A, Cremer F, Coupland G (2002) Control of flowering time. interacting pathways as a basis for diversity. Plant Cell 14 Suppl:S111–130

Murashige T, Kamemoto H, Sheehan TJ (1967) Experiments on the seasonal flowering behaviour of *Vanda* Miss Joaquim. Proc Amer Soc Hort Sci 91:672–679

Nemhauser JL, Zambryski PC, Roe JL (1998) Auxin signaling in *Arabidopsis* flower development? Commentary. Curr Opin Plant Biol 1:531–535

Newton LA, Runkle ES (2009) High-temperature inhibition of flowering of *Phalaenopsis* and *Doritaenopsis* orchids. HortScience 44:1271–1276

Newton LA, Runkle ES (2015) Effects of benzyladenine on vegetative growth and flowering of potted *Miltoniopsis* orchids. Acta Hortic 1078:121–127

Nisha N, Siang T, Maziah M (2012) Effect of 6-benzylaminopurine on flowering of a *Dendrobium* orchid. Aus J Crop Sci 6:225–231

Ohno H (1991) Microsporogenesis and flower bud blasting as affected by high temperature and gibberelic acid in *Cymbidium* (Orchidaceae). J Jpn Soc Hortic Sci 60:149–257

Panjama K, Suzuki E, Otani M, Nakano M, Ohtake N, Ohyama T, Bundithya W, Sueyoshi K, Ruamrungsri S (2019) Isolation and functional analysis of FLOWERING LOCUS T orthologous gene from *Vanda* hybrid. J Plant Biochem Biotechnol 28:374–381

Parcy F, Nilsson O, Busch MA, Lee I, Weigel D (1998) A genetic framework for floral patterning. Nature 395:561–566

Perilli S, Moubayidin L, Sabatini S (2010) The molecular basis of cytokinin function. Curr Opin Plant Biol 13:21–26

Pillitteri LJ, Lovatt CJ, Walling LL (2004) Isolation and characterization of a TERMINAL FLOWER homolog and its correlation with juvenility in citrus. Plant Physiol 135:1540–1551

Poethig RS (2010) The past, present, and future of vegetative phase change. Plant Physiol 154:541–544

Powell C, Caldwell KI, Littler RA, Warrington I (1988) Effect of temperature regime and nitrogen fertilizer level on vegetative and reproductive bud development in *Cymbidium* orchids. J Amer Soc Hortic Sci 113:552–556

Pridgeon A (2000) The Illustrated Encyclopedia of Orchids. Timber Press, Oregon

Rhoades MW, Reinhart BJ, Lim LP, Burge CB, Bartel B, Bartel DP (2002) Prediction of plant microRNA targets. Cell 110:513–520

Rittershausen B, Rittershausen W (2003) Growing Orchids. Hermes House, London

Rotor GB (1952) Daylength and temperature in relation to growth and flowering of orchids. Agr Exp Stn Bull 885:3–47

Rotor GB, Withner CL (1959) The photoperiodic and temperature response of orchids. In: The Orchids: a Scientific Survey, NY, USA, p 397–416, Ronald Press

Sakai WS, Adams C, Braun G (2000) Pseudobulb injected growth regulators as aids for year around production of Hawaiian *Dendrobium* orchid cutflowers. Acta Hortic 541:215–220

Sakanishi Y, Imanishi H, Ishida G (1980) Effect of temperature on growth and flowering of *Phalaenopsis amabilis*. Bull Univ Osaka Ser B: Agr Biol 32:1–9

Sanford WW (1974) The ecology of orchids. Withner, CL (ed) The Orchids: Scientific Studies. John Wiley and Sons, New York, pp 123–132

Sawettalake N, Bunnag S, Wang Y, Shen L, Yu H (2017) DOAP1 promotes flowering in the orchid *Dendrobium* Chao Praya Smile. Front Plant Sci 8:400

Searle I, Coupland G (2004) Induction of flowering by seasonal changes in photoperiod. EMBO J 23:1217–1222

Sheehan TJ (1983) Recent advances in botany, propagation, and physiology of orchids. Hortic Rev 5:279–315

Simpson GG, Dean C (2002) Arabidopsis, the rosetta stone of flowering time? Science 296:285–289

Sinoda K, Suto K, Hara M, Aoki M (1988) Effect of day and night temperature on the flowering of *Dendrobium nobile*-type cultivars. Bull Natl Res Inst Veg Ornamental Plants Tea Ser A 2279290

Song YH, Ito S, Imaizumi T (2013) Flowering time regulation: photoperiod- and temperature-sensing in leaves. Trends Plant Sci 18:575–583

Su WR, Chen WS, Koshioka M, Mander LN, Hung LS, Chen WH, Fu YM, Huang KL (2001) Changes in gibberellin levels in the flowering shoot of *Phalaenopsis hybrida* under high temperature conditions when flower development is blocked. Plant Physiol Biochem 39:45–50

Sun X, Qin Q, Zhang J, Zhang C, Zhou M, Paek KY, Cui Y (2012) Isolation and characterization of the FVE gene of a *Doritaenopsis* hybrid involved in the regulation of flowering. Plant Growth Regul 68:77–86

Sun Y, Wang G, Li Y, Jiang L, Yang Y, Guan S (2016) De novo transcriptome sequencing and comparative analysis to discover genes related to floral development in *Cymbidium faberi* Rolfe. SpringerPlus 5:1458–1458

Teoh ES (2005) Orchids of Asia. Marshall Cavendish Editions, Singapore, p 367

Trevaskis B, Hemming MN, Dennis ES, Peacock WJ (2007) The molecular basis of vernalization-induced flowering in cereals. Trends Plant Sci 12:352–357

Vaz APA, Figueiredo-Ribeiro RdCL, Kerbauy GB (2004) Photoperiod and temperature effects on in vitro growth and flowering of *P. pusilla*, an epiphytic orchid. Plant Physiol Biochem 42:411–415

Wang YT (1995) *Phalaenopsis* orchid light requirement during the induction of spiking. HortScience 30:59–61

Wang SL, ViswanathKK Tong CG, An HR, Jang S, Chen FC (2019) Floral induction and flower development of orchids. Front Plant Sci 10:1258

Weigel D, Alvarez J, Smyth DR, Yanofsky MF, Meyerowitz EM (1992) LEAFY controls floral meristem identity in *Arabidopsis*. Cell 69:843–859

Wee SH (1971) Maturation period of pods and time taken for plant to flower. Malay Orchid Rev 10:42–46

Wen Z, Guo W, Li J, Lin H, He C, Liu Y, Zhang Q, Liu W (2017) Comparative transcriptomic analysis of vernalization- and cytokinin-induced floral transition in *Dendrobium nobile*. Sci Rep 7:45748

Went FW (1957) The experimental control of plant. Chron Bot 17:148–152

Werner T, Schmülling T (2009) Cytokinin action in plant development. Curr Opin Plant Biol 12:527–538

Weyers JDB, Paterson NW (2001) Plant hormones and the control of physiological processes. New Phytol 152:375–407

Wigge PA, Kim MC, Jaeger KE, Busch W, Schmid M, Lohmann JU, Weigel D (2005) Integration of spatial and temporal information during floral induction in *Arabidopsis*. Science 309:1056–1059

Williams BS, Williams H (1894) The Orchid Growers' Manual, 7th ed. Victoria and Paradise Nurseries, London. [Reprinted 1961 by Wheldon and Wesley, Codicote, Hitchin, England, and Hafner Publishing, New York

Wilson RN, Heckman JW, Somerville CR (1992) Gibberellin is required for flowering in *Arabidopsis thaliana* under short days. Plant Physiol 100:403–408

Wu PH, Chang DCN (2009) The use of N-6-benzyladenine to regulate flowering of *Phalaenopsis* orchids. HortTechnology 19:200–203

Xiang L, Li X, Qin D, Guo F, Wu C, Miao L, Sun C (2012) Functional analysis of FLOWERING LOCUS T orthologs from spring orchid (*Cymbidium goeringii* Rchb. f.) that regulates the vegetative to reproductive transition. Plant Physiol Biochem 58:98–105

Xu YL, Gage DA, Zeevaart J (1997) Gibberellins and stem growth in *Arabidopsis thaliana* (Effects of photoperiod on expression of the GA4 and GA5 Loci). Plant Physiol 114:1471–1476

Yan L, Wang X, Liu H, Tian Y, Lian J, Yang R, Hao S, Wang X, Yang S, Li Q, Qi S, Kui L, Okpekum M, Ma X, Zhang J, Ding Z, Zhang G, Wang W, Dong Y, Sheng J (2015) The genome of *Dendrobium officinale* Illuminates the biology of the important traditional Chinese orchid herb. Mol Plant 8:922–934

Yang F, Zhu G, Wei Y, Gao J, Liang G, Peng L, Lu C, Jin J (2019) Low-temperature-induced changes in the transcriptome reveal a major role of CgSVP genes in regulating flowering of *Cymbidium goeringii*. BMC Genom 20:53

Yang T, He Y, Niu S, Yan S, Zhang Y (2020) Identification and characterization of the CONSTANS (CO)/CONSTANS-like (COL) genes related to photoperiodic signaling and flowering in tomato. Plant Sci 301:110653

Yanovsky MJ, Kay SA (2003) Living by the calendar: how plants know when to flower. Nat Rev Mol Cell Biol 4:265–276

Yen, CYT (2008) Effects of nutrient supply and cooling on growth, flower bud differentiation, and propagation of the nobile *Dendrobium* orchid. MS thesis Texas A&M University College Station, TX

Yen CYT, Starman TW, Wang YT, Nie N (2008) Effects of cooling temperature and duration on flowering of the nobile *Dendrobium* orchid. HortScience 43:1765–1769

Yoneda K, Momose H (1990) Effects on flowering of *Phalaenopsis* caused by spraying growth regulators when transferred to highlands. Bull Coll Agric Vet Med Nihon Univ 47:71–74

Yoneda K, Momose H, Kubota S (1991) Effects of daylength and temperature on flowering in juvenile and adult *Phalaenopsis* plants. J Jpn Soc Hortic Sci 60:651–657

Yu H, Goh CJ (2000) Identification and characterization of three orchid MADS-box genes of the AP1/AGL9 subfamily during floral transition. Plant Physiol 123:1325–1336

Yu H, Yang SH, Goh CJ (2000) DOH1, a class 1 knox gene, is required for maintenance of the basic plant architecture and floral transition in orchid. Plant Cell 12:2143–2159

Zaharah H, Saharan HA, Nuraini I (1986) Some experiences with BAP as a flower inducing hormone. Malaysian Orchid Bull 3:31–38

Zhang GQ, Liu KW, Li Z, Lohaus R, Hsiao YY, Niu SC, Wang JY, Lin YC, Xu Q, Chen LJ, Yoshida K, Fujiwara S, Wang ZW, Zhang YQ, Mitsuda N, Wang M, Liu GH, Pecoraro L, Huang HX, Xiao XJ, Lin M, Wu XY, Wu WL, Chen YY, Chang SB, Sakamoto S, Ohme Takagi M, Yagi M, Zeng SJ, Shen CY, Yeh CM, Luo YB, Tsai WC, Van de Peer Y, Liu ZJ (2017) The *Apostasia* genome and the evolution of orchids. Nature 549:379–383

Zhang GQ, Xu Q, Bian C, Tsai WC, Yeh CM, Liu KW, Yoshida K, Zhang LS, Chang SB, Chen F, Shi Y, Su YY, Zhang YQ, Chen LJ, Yin Y, Lin M, Huang H, Deng H, Wang ZW, Zhu SL, Zhao X, Deng C, Niu SC, Huang J, Wang M, Liu GH, Yang HJ, Xiao XJ, Hsiao YY, Wu WL, Chen YY, Mitsuda N, OhmeTakagi M, Luo YB, Van de Peer Y, Liu ZJ (2016) The *Dendrobium catenatum* Lindl. genome sequence provides insights into polysaccharide synthase, floral development and adaptive evolution. Sci Rep 6:19029

Zhang JX, Wu KL, Tian LN, Zeng SJ, Duan J (2011) Cloning and characterization of a novel CONSTANS-like gene from *Phalaenopsis hybrida*. Acta Physiol Plant 33:409–417

Zhang JX, Wu KL, Zeng SJ, Duan J, Tian LN (2010) Characterization and expression analysis of PhalLFY, a homologue in *Phalaenopsis* of FLORICAULA/LEAFY genes. Sci Hortic 124:482–489

Zhang J, Wu K, Zeng S, Teixeira da Silva JA, Zhao X, Tian C-E, Xia H, Duan J (2013) Transcriptome analysis of *Cymbidium sinense* and its application to the identification of genes associated with floral development. BMC Genom 14:279

Zhou S, Jiang L, Guan S, Gao Y, Gao Q, Wang G, Duan K (2018) Expression profiles of five FT-like genes and functional analysis of PhFT-1 in a *Phalaenopsis* hybrid. Electron J Biotechnol 31:75–83

Zwanenburg B, Blanco-Ania D (2018) Strigolactones: new plant hormones in the spotlight. J Exp Bot 69:22-5-2218

The Roles of MADS-Box Genes During Orchid Floral Development

7

Jian-Zhi Huang, Pablo Bolaños-Villegas, I-Chun Pan, and Fure-Chyi Chen

Abstract

Orchids have flowers of unique beauty that are remarkable for their zygomorphic syndrome, which can be summarized as a floral architecture based on three categories of organs at the perianth: external tepals, internal lateral tepals, and a labellum or lip, a prominent central inner petal believed to be a specialized adaptation that attracts pollinators. These mesmerizing floral traits have enthralled researchers into the study of the orchid homologs of the MADS-box family of genes, which are transcriptional factors believed to be spatiotemporal determinants of organ identity during floral development. The identification of several putative members of the MADS-box family may clarify their potential role in orchid flower development, especially during the transition to flowering and during the patterning of orchid flower organs. Furthermore, we look into new technologies of genome analysis and gene editing in order to appraise potential applications for basic research purposes and for the breeding of new orchid varieties.

J.-Z. Huang · F.-C. Chen (✉)
Department of Plant Industry, National Pingtung University of Science and Technology, Pingtung, Taiwan
e-mail: fchen@mail.npust.edu.tw

J.-Z. Huang
Department of Agricultural Chemistry, College of Bioresources and Agriculture, National Taiwan University, Taipei, Taiwan

P. Bolaños-Villegas
Fabio Baudrit Agricultural Research Station, University of Costa Rica, La Garita, Alajuela 20101, Costa Rica

Jardín Botánico Lankester, Universidad de Costa Rica, Cartago P.O. Box 302-7050 Costa Rica

I.-C. Pan
Department of Horticulture, National Chung Hsing University, Taichung, Taiwan

7.1 Introduction

By most accounts, the Orchidaceae has been considered as one of the largest family of plants within the angiosperms, with perhaps as much as 25,000 species scattered across the planet (Leitch et al. 2009; Petruzzello 2018). Orchids are cultivated commercially either as pot plants or for the production of beautiful cut flowers (Griesbach et al. 2002), especially in tropical and subtropical countries. Orchids from genus *Phalaenopsis* are currently one of the most relevant flower commodities since they can be cultivated year round thanks to control of flowering by temperature. This genus comprises of approximately 66 natural species according to the latest classification (Christenson 2001). Flowers in *Phalaenopsis* exhibit bilateral symmetry with floral organs that are unequal in size, also known

F.-C. Chen and S.-W. Chin (eds.), *The Orchid Genome*, Compendium of Plant Genomes,
https://doi.org/10.1007/978-3-030-66826-6_7

Fig. 7.1 Flower of *Phalaenopsis* Little Gem Stripes and the illustration of its morphology

as zygomorphic architecture or zygomorphic morphology. These flowers showcase three sepals within the first whorl, but in the second whorl flowers develop two petals plus a distinctive central inner tepal called the labellum (Huang et al. 2015; Fig. 7.1). The gynostemium or (column) is another specialized reproductive organ that consists of the fusion of the original female and male organs. In pollen mother cells, meiosis takes place during flower development, while female meiosis occurs certain times only after pollination (O'Neill 1997). After telophase II, tetrads remain together, pollen mitosis takes place and the resulting mature pollen grains aggregate as easily recognizable pollinium located at the tip of the column (Rudall and Bateman 2002; Aceto and Gaudio 2011; Rajkumari and Longjam 2005).

Flower development in higher plants has been investigated in detail, and the overall consensus is that the process is primarily regulated by a family of homeotic transcriptional factors that feature a conserved sequence motif (60 amino acids) for binding DNA, called the MADS-box motif, and that these MADS-box genes specify floral organ identity and development (Weigel and Meyerowitz 1994; Purugganan et al. 1995; Münster et al. 1997). Based on the existence of five major groups of plant MADS-box genes (A, B, C, D and E) a model, aptly named as the ABCDE model was proposed to explain the process that determines organ identity in flowers (Theißen 2001).

In this chapter, we review recent research on the most current functional characterizations of MADS-box proteins in orchids and how they precisely regulate floral organ development.

7.2 The MADS-Box Family of Transcription Factors

The MADS-box family of genes is believed to consist of transcription factors (TFs) that feature an evolutionarily conserved sequence of about 58 ~ 60 amino acids referred to as the eukaryotic MADS domain. The first MADS-box gene to be isolated was *ARG80* from budding yeast (Passmore et al. 1988). The term MADS-box is a handy acronym for the conceptual amalgamation of homeotic genes *MINICHROMOSOME MAINTENANCE 1* (*MCM1*) from *Saccharomyces cerevisiae* (budding yeast), *AGAMOUS* (*AG*) from *Arabidopsis thaliana* (thale cress) (Yanofsky et al. 1990), *DEFICIENS* (*DEFA*) from *Antirrhinum majus* (snapdragon) (Schwarz-Sommer et al. 1990), and *SERUM RESPONSE FACTOR* (*SRF*) from *Homo sapiens* (Norman et al. 1988). The family of MADS-box genes can be binned into types I and II based on their phylogenetic relatedness (Alvarez-Buylla et al. 2000). Type I genes, also named as the M-type, contain a conserved M domain and show hypervariability at the carboxy terminus, these genes further comprise three subgroups called Mα, Mβ, and Mγ (Masiero et al. 2011). Type II genes are believed to encode MIKC proteins, thus named for featuring a MADS (M) domain, an intervening (I) domain, a keratin-like (K) domain, and a characteristic C-terminal (C) domain (Kaufmann et al. 2005). All these TFs contain a characteristic amino-terminal domain (the MADS domain) of 60 residues that bind to target genes at a regulatory region that features a CC $(A/T)_6$GG sequence called the CArG-box (Messenguyand Dubois 2003; Riechmann et al. 1996; Hayes et al. 1988). These MADS-box genes are believed to be homeotic, and they specify organ identity and, flowering time and several other reproductive processes (Dornelas et al. 2011). These genes are also expressed in somatic tissues but their function is not well understood (Messenguy and Dubois 2003; Pařenicová et al. 2003). In plants, floral development is hypothesized to operate in accordance to ABCDE model, in which five types of homeotic factors specify development (A–E), most of which appear to be MADS-box genes (Theißen 2001). Genes from the A and E classes exert developmental control in sepals located within the first whorl (Ditta et al. 2004). Formation of petals within the second whorl is jointly regulated by the A, B and E classes, while the development of stamens within the third whorl is under concerted control of the B, C and E classes. Development of carpel in the fourth whorl is believed to be under shared control by genes from the C and E classes. Finally, ovule development is thought to be the sole responsibility of genes from the D-class (Theißen and Melzer 2007; Krizek and Fletcher 2005). In the model plant, *Arabidopsis thaliana*, some notable genes from the A-class are *APETALA1* (*AP1*) and *APETALA2* (*AP2*) (Mandel and Yanofsky 1995; Jofuku et al. 1994). The B-class is represented by genes *AP3* and *PISTILLATA* (*PI*) (Goto and Meyerowitz 1994; Jack et al. 1992), the C-class by gene *AGAMOUS* (*AG*) (Pinyopich et al. 2003), the D-class by gene *SEEDSTICK/AGAMOUS-LIKE11* (*STK/AGL11*) (Pinyopich et al. 2003) and the E-class is represented by the *SEPALLATA* genes (*SEP1*, *SEP2*, *SEP3*, and *SEP4*) (Pelaz et al. 2000; Ditta et al. 2004).

7.3 Molecular Determination of Organ Development in Orchid Flowers

As mentioned earlier, orchid flowers show a distinctive zygomorphic structure, composed of two whorls of tepals, a central gynostemium, three sepals and three petals, one of which is a very prominent, fleshy, highly modified labellum or lip located just below the gynostemium because of resupination (Rudall and Bateman 2002; Mondragón-Palomino and Theißen 2009). The delicate genetic determination of such a sophisticated floral organization as an adaptation to pollinators is believed to be a key factor behind the prodigious diversification of the Orchidaceae family (Mondragón-Palomino and Theißen 2009). For instance, the orchid B-class MADS-box genes, *PhAGL6a* and *PhAGL6b*, show a very specific expression pattern in the labellum of flower from genus *Phalaenopsis*, and may operate as positive regulators of its formation (Su et al. 2013b; Huang et al. 2015).

Putative genes from the A-class have been successfully identified in several orchid genera. For example, in genus *Phalaenopsis,* transcripts related to the *Arabidopsis SQUAMOSA* (*SQUA*) family of genes, including *ORAP11* and *ORAP13*, have been isolated from flower bud tissue (Chen et al. 2007). Other examples are *OMADS10* from *Oncidium* (Chang et al. 2009); and *DOMADS2*, and *DthyrFL1/2/3* from *Dendrobium* (Yu and Goh 2000; Skipper et al. 2006). Ectopic expression of these A-class genes in transgenic plants may induce early flowering and changes in plant architecture (Chang et al. 2009; Chen et al. 2007; Skipper et al. 2006; Yu and Goh 2000). In *Oncidium*, the gene *OMADS3* is believed to be an ancestral *AP3*-like gene that operates as an A-class regulator of floral initiation and formation (Hsu and Yang 2002). In addition, three more ancestral A-class *AP3-like* genes, namely *OMADS5*, *OMADS3* and

OMADS9, and one *PISTILLATA-like* gene, *OMADS8*, have been identified and characterized in *Oncidium* (Chang et al. 2010). As for the B-class, in *Phalaenopsis*, it was possible to determine that *DEF*-like transcripts showing an altered expression pattern in peloric mutants with distinctive lip-like petals (Tsai et al. 2004; Mondragón-Palomino and Theißen 2011). In the genus *Habenaria* (the bog orchids), several putative B-class genes such as *HrGLO1* and *HrGLO2* were found to be highly expressed in sepals, petals and in the column, but *HrDEF* only showed expression in the petals and column (Kim et al. 2007). Prospective genes from the C and D-classes have been successfully identified in four orchid genera: *PhalAG1* and *PhalAG2* from *Phalaenopsis* (Song et al. 2006), *DthyrAG1*, *DthyrAG2* (Skipper et al. 2006), *DcOAG1* and *DcOAG2* (Xu et al. 2006) from *Dendrobium* and two putative *AGAMOUS*-like genes named *CeMADS1* and *CeMADS2* from *Cymbidium* (Wang et al. 2011). Possible genes from the E-class related to *AGL6* have been identified in *Oncidium*, these are: *OMADS11*, which was assigned to the *LOFSEP* subclass, *OMADS6* from the *SEP3* subclass, and *OMADS1* and *OMADS7* from the *AGL6* subclass (Hsu et al. 2003; Chang et al. 2009).

Based on the expression pattern and their hypothetical role in development, we propose a model for how these putative MADS-box genes determine organ identity in orchid flowers (see below). We believe that it is necessary to sketch possible relationship networks in order to improve our current knowledge about the function of orchid MADS-box genes, as suggested by others (Theißen et al. 2016).

7.4 The ABCDE Model in Orchids

7.4.1 A-Class Genes

Evidence drawn from *Arabidopsis thaliana* and *Antirrhinum majus*, suggests that A-class genes, such as *Arabidopsis APETALA1* (*AP1*) and *Antirrhinum SQUAMOSA* (*SQUA*), generally play three main developmental functions: induction of the transition from the vegetative to the reproductive growth phase, determination of organ identity in flowers and the regulation of fruit ripening (Fornara et al. 2004). Genes within the orchid *SQUA-like* subgroup are further divided into two clades related to genes found in the eudicots: the *FUL/AGL8-like* genes and *euAP1-like* genes, which are groups in clear evolutionary divergence with the *AP1* subgroup, which is ancestral and closely related to the monocots (Litt and Irish 2003; Vandenbussche et al. 2003a). In Orchidaceae, several *euAP1-like* genes have been identified in genera *Cymbidium*, *Dendrobium*, *Oncidium*, *Phalaenopsis* and the terrestrial genus *Orchis* (Yang and Zhu 2015; Yang et al. 2019; Hsu et al. 2015; Valoroso et al. 2019; Chang et al. 2009; Chen et al. 2007; Yu and Goh 2000).

In *Dendrobium*, three A-class genes (*DOMADS1/2/3*) are expressed in its shoot apical meristem during the floral transition and in mature flowers. Transcripts from *DOMADS1* are expressed uniformly in the flower meristem, the floral primordium and possibly in mature flower organs. *DOMADS2* is expressed specifically in the column and ovaries and the expression of *DOMADS3* is restricted to the pedicel (Yu and Goh 2000). In *Arabidopsis*, the expression of *AP1* is restricted to the sepals and petals, therefore the differences observed in *Dendrobium* may imply the existence of functional divergences in these *AP1*-like genes (Sawettalake et al. 2017, 2000, 2000). In the hybrid *Dendrobium* Chao Praya Smile, *DOAP1* is highly expressed in all reproductive tissues, particularly in the flower apices. Ectopic overexpression of *DOAP1* in wild type *Arabidopsis* and *Dendrobium* results in early flowering phenotypes, and rescues organ developmental defects in flowers of the *Arabidopsis ap1* mutant (Sawettalake et al. 2017).

OMADS10 is a putative ancestral *AP1* ortholog identified from tissues of *Oncidium* (now *Oncidesa*) Gower Ramsey. *OMADS10* is only expressed in leaves and in the lip and carpels of mature flowers, which is a pattern that differs from what has been reported for other genes in the *SQUA* subfamily. Most A-class genes from the *SQUA-like* subfamily conform to an expression pattern restricted to the early floral meristem, and expression is absent in vegetative leaves. Ectopic expression of the *OMADS10* transcript in *Arabidopsis* induces early flowering without causing any developmental defects in floral organs (Chang et al. 2009). A similar situation has been observed with genes *ORAP11* and *ORAP13* from *Phalaenopsis hybrida* cv. Formosa Rose. Both are strongly expressed in floral buds but are also detected in vegetative tissues. In *Nicotiana tabacum*, overexpression of both genes induces early flowering and changes in plant architecture. Thus, *ORAP11* and *ORAP13* have evolved novel partially overlapping roles both in the transition to flowering and during the determination of whole plant architecture (Chen et al. 2007).

In the terrestrial naked man orchid (*Orchis italica*), four *AP1/FUL* transcripts (*OIcomp2508_AP1*, *OIcomp3679_AP1*, *OIcomp9283_AP1*, *OIcomp11046_AP1*) are differentially expressed in the inflorescences. The *AP1/FUL* transcripts *OIcomp3679*, *9283*, *2508* are highly expressed in the column and ovaries, while the *OIcomp11046* transcript is highly expressed in the outer and inner tepals, and comparatively less in the column and ovaries (Valoroso et al. 2019). Remarkably, this expression pattern is very similar to what has been reported in the twig epiphyte *Erycina pusilla* for genes *EpMADS10/11/12* (Lin et al. 2016). These studies suggest that the orchid orthologs for *AP1* have retained the same core functions during the promotion of the floral transition and during the determination of floral organ development.

7.4.2 B-Class Genes

In *Arabidopsis*, the determination of petal and stamen identity is controlled by MADS-box genes from the B-class (Fornara et al. 2003). Two of such B-class genes are *AP3* (*APETALA3*) and *PI* (*PISTILLATA*), which are functionally analogous to the *DEF* (*DEFICIENS*) and *GLO* (*GLOBOSA*) from snapdragon (Zahn et al. 2006; Winter et al. 2002; Becker and Theißen 2003). Evidence indicates that the function of B-class genes in monocots such as rice, maize and lily and in eudicots have remained comparatively similar (Ambrose et al. 2000; Tzeng and Yang 2001; Lee et al. 2003; Nagasawa et al. 2003; Whipple et al. 2004). As seen on Table 7.1, the molecular function of orchid B-class genes has drawn lot of attention and led to a level of functional characterization unmatched within homeotic genes (Tsai et al. 2005; Mondragón-Palomino and Theißen 2008; Tsai et al. 2014; Mao et al. 2015; Huang et al. 2016; Chang et al. 2009, 2010; Hsu et al. 2015; Hsu and Yang 2002; Huang et al. 2015; Su et al. 2013b; Tsai et al. 2004; Xu et al. 2006; Yang and Zhu 2015).

Many orchid B-class genes show an expression pattern that spills out of the first floral whorl, probably suggesting a role for the developmental differentiation of petaloid sepals. For instance, in *P. equestris*, there are four *DEF*-like genes named *PeMADS2/3/4/5*, which belong to the *AP3* phylogenetic lineage, and *PeMADS6* which may belong to the distinct *PI* lineage. Their expression pattern in floral organs is notably different. Genes *PeMADS2* and *PeMADS5* from the AP3 lineage are expressed in sepals, petals, lip and column of wild-type flowers, but only *PeMADS2* is detected in the sepal of both wild type and peloric flowers (e.g. with lip-like petals), suggesting high functional specificity for sepal development. The expression of *PeMADS5* is apparently below the detection threshold, and no function can be deduced, whereas the *PeMADS4* transcript is detected in the lip-like petals of the peloric mutant, suggesting that *PeMADS4* expression is required for proper lip development. In addition, chromatin harvested from lip tissues is differentially enriched in permissive histone acetylation marks (H3K/9K14ac) located precisely at the translation initiation site of *PeMADS4*. Thus, it is possible that during the development of *Phalaenopsis* floral primordia, the high expression level of *AP3/DEF-like* genes may be brought about by the action of upstream cis-elements being recruited by histone acetylation marks (Hsu et al. 2014). Taken together this body

Table 7.1 A summary of MADS-box regulators in orchids

Orchid	Orchid orthologs	Class	References
Phalaenopsis hybrida cv. Formosa Rose	ORAP11, ORAP13	A	Chen et al. (2007)
Phalaenopsis aphrodite	PaAP1-1, PaAP1-2, PaAP2-5, PaAP2-7, PaAP2-11		Su et al. (2013b)
Phalaenopsis Athens	PhaMADS1, PhaMADS2		Acri-Nunes-Miranda and Mondragon-Palomino (2014)
Dendrobium Madam Thong-in	DOMADS2		Yu et al. (2000)
Dendrobium thyrsiflorum	DthyrFL1, DthyrFL2, DthyrFL3		Skipper et al. (2006)
Dendrobium Chao Praya Smile	DOAP1		Sawettalake et al. (2017)
Cymbidium ensifolium	ZHLZ.comp57026		Yang and Zhu (2015)
Cymbidium faberi	CfAP11		Tian et al. (2013)
Oncidium Gower Ramsey	OMADS10 (OAP1)		Chang et al. (2009), Hsu et al. (2015)
Orchis italica	Olcomp2508_AP1, Olcomp3679_AP1, Olcomp9283_AP1, Olcomp11046_AP1		Valoroso et al. (2019)
Cymbidium goeringii	CgAP1		Yang et al. (2019)
Cymbidium ensifolium	CeAP3, ZHLH.comp53790, ZHLZ.comp35346, ZHLZ.comp55590, ZHLZ.	B	Yang and Zhu (2015)
Cymbidium hybrid cultivar	MADS1		Aceto and Gaudio (2011)
Dendrobium crumenatum	DcOAP3A; DcOAP3B		Xu et al. (2006)
Dendrobium moniliforme	DMMADS4		Aceto and Gaudio (2011)
Gongora galeata	GogalDEF1, GogalDEF2, GogalDEF3		Aceto and Gaudio (2011)
Habenaria radiata	HrDEF		Aceto and Gaudio (2011)
Oncidium Gower Ramsey	OMADS3 (OAP3-3),OMADS5 (OAP3-1), OMADS9 (OAP3-2)		Hsu and Yang (2002), Hsu et al. (2015)
Orchis italica	Olcomp900_DEF4, Olcomp3831_DEF1, Olcomp7668_DEF3, Olcomp22604_DEF2		Valoroso et al. (2019)
Phalaenopsis equestris	PeMADS2, PeMADS3 PeMADS4, PeMADS5		Tsai et al. (2004), Hsieh et al. (2013)
Phragmipedium longifolium	PhlonDEF1, PhlonDEF2, PhlonDEF3, PhlonDEF4		Aceto and Gaudio (2011)
Spiranthes odorata	SpodoDEF1, SpodoDEF2, SpodoDEF3		Aceto and Gaudio (2011)
Vanilla planifolia	VaplaDEF1, VaplaDEF2, VaplaDEF3		Aceto and Gaudio (2011)
Phalaenopsis aphrodite	PaAP3-1, PaAP3-2, PaAP3-3, -4		Su et al. (2013b)
Phalaenopsis equestris	PeMADS1	C	Chen et al. (2012)
Phalaenopsis Hatsuyuki	PhalAG1		Song et al. (2006)
Phalaenopsis Athens	PhaMADS8, PhaMADS10		Acri-Nunes-Miranda and Mondragon-Palomino (2014)

(continued)

Table 7.1 (continued)

Orchid	Orchid orthologs	Class	References
Dendrobium crumenatum	DcOAG1		Skipper et al. (2006)
Dendrobium thyrsiflorum	DthyrAG1		Xu et al. (2006)
Oncidium Gower Ramsey	OMADS4		Hsu et al. (2010)
Cymbidium ensifolium	CeMADS1, CeMADS2		Wang et al. (2011)
Orchis italica	OitaAG		Salemme et al. (2013)
Erycina pusilla	EpMADS20, EpMADS21, EpMADS22		Dirks-Mulder et al. (2017)
Phalaenopsis equestris	PeMADS7	D	Chen et al. (2012)
Phalaenopsis Hatsuyuki	PhalAG2		Song et al. (2006)
Phalaenopsis Athens	PhaMADS9		Acri-Nunes-Miranda and Mondragon-Palomino (2014)
Dendrobium crumenatum	DcOAG2		Skipper et al. (2006)
Dendrobium thyrsiflorum	DthyrAG2		Xu et al. (2006)
Oncidium Gower Ramsey	OMADS2		Hsu et al. (2010)
Cymbidium ensifolium	CeMADS1		Wang et al. (2011)
Orchis italica	OitaSTK		Salemme et al. (2013)
Erycina pusilla	EpMADS23		Dirks-Mulder et al. (2017)
Aranda Deborah	OM1	E	Lu et al. (1993)
Phalaenopsis Athens	PhaMADS4, PhaMADS5, PhaMADS7		Acri-Nunes-Miranda and Mondragon-Palomino (2014)
Phalaenopsis equestris	PeSEP1, PeSEP2, PeSEP3, PeSEP4		Pan et al. (2014)
Dendrobium Madam Thong-in	DOMADS1, DOMADS3		Yu and Goh (2000)
Dendrobium crumenatum	DcOSEP1		Xu et al. (2006)
Oncidium Gower Ramsey	OMADS6, OMADS10, OMADS11		Chang et al. (2009)
Cymbidium goeringii	CgSEP1, CgSEP2, CgSEP3, CgSEP4		Xiang et al. (2018)
Habenaria radiata	HrSEP-1		Mitoma and Kanno (2018)
Phalaenopsis aphrodite	PaAGL6-1, PaAGL6-2	AGL6-like	Su et al. (2013b)
Phalaenopsis Brother Spring Dancer 'KHM190'	PhAGL6a, PhAGL6b		Huang et al. (2015), Huang et al. (2016)
Cymbidium goeringii	CgAGL6-1, CgAGL6-2, CgAGL6-3		Xiang et al. (2018)
Oncidium Gower Ramsey	OMADS1, OMADS7		Chang et al. (2009), Hsu et al. (2015)

of evidence supports the notion that the *Phalaenopsis* genes *PeMADS2*/3/4 play a decisive role in the development of flower organs, including the lip (Tsai et al. 2004, 2005).

In the pigeon orchid *Dendrobium crumenatum*, the B-class genes *DcOAP3A* and *DcOPI* are expressed in the tissues of all floral organs, whereas the expression of closely related gene *DcOAP3B* is restricted to the second whorl of petals, the lip, the column and the pollinia, but it is not clear why. A complementation test in the *Arabidopsis pi-1* null mutant using the *DcOPI* open reading frame rescued all evident flowering developmental defects, suggesting that function has been retained through evolution. Moreover, results from yeast two-hybrid assays suggest that *DcOAP3A/B* and *DcOPI* can form heterodimer modules (Xu et al. 2006), but the reason is not clear.

In *Oncidium* Gower Ramsey, it was possible to identify three presumed ancestral *APETALA3-like* genes: *OMADS3/5* and *OMADS9*, and one *PI*-like gene, *OMADS8* (Chang et al. 2010; Hsu and Yang 2002). Tissue characterization in mature flowers indicates that in the *AP3*-like group, *OMADS3* and *OMADS8* are expressed in all four whorls, but unexpectedly in leaves as well, while *OMADS5* is highly expressed in sepals and petals but not in the lip, and it is significantly down-regulated in the lip-like petals and sepal of peloric mutant flowers (Chang et al. 2010). The *PI*-like gene *OMADS9* is expressed in the petals and lip. Yeast two-hybrid assays showed that OMADS3, OMADS5 and OMADS9 can form homodimers; but in addition, heterodimers may be formed between OMADS3 andOMADS8, and between OMADS5 and OMADS9. Thus, in *Oncidium*, the formation of sepals, petals and lip may require the presence of stable transcriptional complexes between OMADS3/8 or by a OMADS9 homodimer. Nonetheless, the determination of final organ identity in the sepals, petals and lip may depend on the presence of OMADS5, as suggested from peloric flowers (Chang et al. 2010).

In *Habenaria radiata* (the Korean Sagiso orchid), three B-class MADS-box genes have been identified, including two *PI/GLO*-like genes named *HrGLO1*, *HrGLO2* and one *AP3/DEF-like* gene, predictably named *HrDEF* (Kim et al. 2007). Genes *HrGLO1* and *HrGLO2* are conspicuously expressed in sepals, petals and in the column, whereas *HrDEF* expression is only detected in the petals and in the column, but not in sepals. All these three B-class genes are expressed in petaloid-sepals, petals and in the column of a peloric mutant with petaloid-sepals (Kim et al. 2007). This result is believed to support the hypothesis that changes in the expression of *HrDEF* define organ identity in sepals and petals of *H. radiata* (Kim et al. 2007).

A mechanism for determination of floral development in *Cymbidium goeringii* Rchb (the noble orchid) was deduced from transcriptome expression analysis of B-class genes along with gene co-expression network analysis involving the A and E classes (Xiang et al. 2018, 2018). Results indicate that genes *CgDEF1*, *CgSEP2* and *CgAGL6-1* are highly upregulated in the sepals and petals and are visibly down-regulated in the lip. In contrast, *CgDEF3*, *CgDEF4* and *CgAGL6-3* are highly expressed in the lip and in peloric lip-like petals. Biochemical characterization by yeast two-hybrid assays suggests that the proteins CgDEF1 and CgGLO can form heterodimers. In addition, CgAGL6-1/CgSEP2 and CgDEF1 can form heterodimers in the presence of the CgGLO protein, and CgAGL6-1 can form a heterodimer with CgSEP2. It also appears that CgDEF3 and CgDEF4 may interact independently with CgGLO and CgAGL6-3, respectively; while CgDEF3 and CgDEF4 may form heterodimers in the presence of CgGLO. Taken together, these results may be interpreted as suggesting that lip formation is specified by a CgDEF3/CgDEF4/CgAGL6-3/CgGLO complex ('the lip quartet'), whereas the CgDEF1/CgAGL6-1/CgSEP2/CgGLO complex ('sepal/petal-quartet') may promote sepal and petal formation instead (Xiang et al. 2018).

Several other putative B-class genes have been identified in species *Orchis italica* (Salemme et al. 2013) and *Erycina pusilla* (Dirks-Mulder et al. 2017). In immature flowers of *Orchis italica*, the presumed *PI/GLO-like* paralogs *OrcPI* and *OrcPI2* are expressed in all organs. In mature flowers, their expression is restricted to the lip, and the expression of *OrcPI2* is the highest of both. In the genome of *E. pusilla*, there are three putative homologs of *AP3* (*EpMADS13*, *EpMADS14*, *EpMADS15*) and a single homolog of *PI* (*EpMADS16*). Their expression is observed in most of the flower organs, nonetheless, *EpMADS14* shows very high and preferential expression in lateral sepals while *EpMADS13* shows high expression in the lip and callus, and comparatively less in sepals and petals (Salemme et al. 2013). The *PI*-like gene *EpMADS16* shows indiscriminate expression across all floral organs (Dirks-Mulder et al. 2017).

When the expression pattern of B-class genes was compared with the developmental pattern of wild type and peloric flowers it was possible to derive a 'modified ABC model' specific for orchids (Mondragón-Palomino and Theißen 2011; Su et al. 2013b).

7.4.3 C- and D-Class Genes

In *Arabidopsis*, genes from the C-class are expressed in whorls 3 and whorl 4, presumably to negatively modulate the expression of A-class genes, and contribute to the formation of the stamen and pistil. In *Arabidopsis* too, the *AGAMOUS* (*AG*) C-class gene is believed to be the sole determinant of male and female organ identity (Bowman et al. 1991; Yanofsky et al. 1990), while in snapdragon, there are two determinant genes, *PLENA* (*PLE*) and *FARINELLI* (*FAR*) (Davies et al. 1999). In snapdragon, the simultaneous reduction in transcription from *AG* and *PLENA* results in the conversion of stamens to petals in the third whorl, and pistils to sepals in the fourth whorl (Davies et al. 1999; Yanofsky et al. 1990). Genes from the D-class have been linked to determination in ovule development (Angenent and Colombo 1996). In petunia, the D-class genes *FLORAL BINDING PROTEIN 7* (*FBP7*) and *FBP11* are considered responsible for the establishment of ovule identity within the carpel (Wittich et al. 1999). Overexpression of the *FBP11* sequence with the CaMV 35S promoter elicits development of ovule-like structures above the sepals and in below the petals (Angenent et al. 1995; Colombo et al. 1995). In contrast, co-suppression of *FBP7* and *FBP11* expression leads to defective development of the seed coat and consequent degeneration of the endosperm, which leads to a shrunken phenotype (Colombo et al. 1997). It is believed that genes from the Cand D classes arose after an event of genome duplication during the evolution of angiosperms (Kramer et al. 2004; Becker and Theißen 2003). All C and D class MADS-box proteins consistently belong to the AG subfamily (Kramer et al. 2004; Theißen et al. 2000), hinting at their common evolutionary origin.

In orchids, the development of column/gynostemium involves the fusion of tissue from whorls 3 and 4, which is one of the most intriguing and challenging subjects yet to fully elucidate regarding the evolution of C-class genes. Also, ovule development in orchid flowers is elicited and completed only as a result of pollination (Swamy 1943; O'Neill 1997). Thus, orchids are useful models for investigating how did D-class genes end up being involved in ovule development.

As shown on Table 7.1, several putative C- and D-class MADS-box genes have been identified in several orchid species. Genes *PeMADS1* and *PeMADS7* from *Phalaenopsis equestris* were characterized as C- and D-class genes, respectively. Analysis of spatial and temporal expression data suggests that they are specifically expressed during development of the gynostemium and ovules. *PeMADS7* is not detected at all flower primordia during early flower development, while *PeMADS1* is significantly overexpressed in petals of the *gylp* (*gynostemium-like petal*) mutant, whose petals were developed into aberrant gynostemium-like structures. This finding suggests that *PeMADS1* fulfills a significant role during gynostemium development (Chen et al. 2012). Protein–protein interaction analysis revealed that neither PeMADS1 nor PeMADS7 can dimerize in the presence of PeMADS8, an E-class protein speculated to operate as a molecular scaffold (Chen et al. 2012). A complementation test showed that ectopic expression of the *PeMADS1* ORF can rescue the phenotype of the *Arabidopsis ag* mutant. In contrast, ectopic expression of *PeMADS7* in *Arabidopsis* causes developmental defects associated with the D-class gene family. In addition, two studies have identified and characterized C- and D-class candidate genes from *Phalaenopsis* cv. Hatsuyuki and *Phalaenopsis* cv. Athens, including several *AG-like* (*PhlAG1*, *PhaMADS8*, *PhaMADS10*) and several *STK-like* ones (*PhlAG2*, *PhaMADS9*) (Acri-Nunes-Miranda and Mondragón-Palomino 2014; Song et al. 2006). The putative *PhlAG1* and *PhlAG2* genes are expressed specifically in the lip, column and ovules, and are not detectable in vegetative organs such as protocorm-like bodies (PLBs), leaves or roots (Song et al. 2006). Genes *PhaMADS8* and *PhaMADS10* are significantly expressed in the gynostemium as in the ovaries, and no apparent pattern is observed for *PhaMADS9* (Acri-Nunes-Miranda and Mondragón-Palomino 2014).

In *Oncidium* Gower Ramsey, the putative *OMADS4* and *OMADS2* genes are candidate C- and D-class factors. *OMADS4* appears to be expressed in flower organs such as the stamens and carpels, while the expression of *OMADS2* is narrow and appears restricted to the stigmatic cavity, ovaries and carpels. Biochemical characterization by the yeast two hybrid assays indicates that OMADS2 and OMADS4 are able to self-dimerize and to form heterodimers with each other. Ectopic overexpression of *OMADS2* and *OMADS4* in *Arabidopsis* causes early flowering but there is no developmental (e.g. homeotic) conversion in any flower organs (Hsu et al. 2010). The C- and D-class genes in *D. crumenatum* and *D. thyrsiflorum* were identified, including *DcOAG1* and *DthyrAG1* as C-class genes; and *DcOAG2* and *DthyrAG2* as D-class genes (Skipper et al. 2006; Xu et al. 2006). *DcOAG1* was expressed in all floral organs, nonetheless, the expression of *DcOAG2* is mostly confined to the base of the column (the ovaries) and in the tissues that surround the pollinia (anther cap and stipe). Interestingly, the ectopic expression of *DcOAG1* in *Arabidopsis* did induce flowering and developmental conversion of sepals to carpel-like structures with stigmatic papillae at the tip (Xu et al. 2006). Besides, the petals were either absent or had transformed into stamen-like structures. These results may be interpreted as suggesting that *DcOAG1* performs similar functions to those of *AG* in *Arabidopsis* (Xu et al. 2006). The functional characterization of *DthyrAG1/2* indicates that *DthyrAG1* is only expressed at the early stages of placenta- and ovule development, but whereas *DthyrAG2* is expressed during the whole process of ovule development. Localization of the *DthyrAG1/2* transcripts by in situ hybridization indicates active transcription in the rostellum, stigma, and the stylar canal. These results lend support to the hypothesis that the orchid C- and D-class homologs are involved in key developmental processes during flower development, especially by *DthyrAG2* during late ovule development, and little less by *DthyrAG1* (Skipper et al. 2006).

In *Cymbidium ensifolium*, the four-season orchid, it was possible to compare the spatial/temporal expression pattern of genes *CeMADS1* and *CeMADS2*. These two putative C-class genes were studied in flower buds from wild type and from multitepal mutant flowers, which are strongly scented. In flowers of the mutant, there is no detectable expression of *CeMADS1*, moreover the column is replaced by an ectopic flower that grows aberrant sepals and petals in a centripetal pattern. In wild type and mutant flowers, *CeMADS2* is expressed in all organs, especially in the column. Overexpression of both genes in *Arabidopsis* reduces the growth of primary flowers. In conclusion, these results suggest that *CeMADS1* may elicit the initial stages of gynostemium development whereas *CeMADS2* allows completion of gynostemium morphogenesis. Results also suggest that the expression of *CeMADS1* is crucial for proper orchid flower morphogenesis (Wang et al. 2011).

Three orchid *AG-like* (*EpMADS20/21/22*) and one *STK-like* genes were identified in the tiny Central American twig epiphyte *Erycina pusilla* (Dirks-Mulder et al. 2017). Data indicate that *EpMADS20* is expressed across all floral organs, while *EpMADS21/22/23* is primarily expressed in the gynostemium (Dirks-Mulder et al. 2017). In *Habenaria radiata*, the putative genes *HrAG-1* and *HrAG-2* are expressed only in the column, which is consistent with the expression pattern expected for *AG-like* orthologs in orchid species (Mitoma and Kanno 2018). In *Orchis italica*, the presumed C- and D-class genes *OitaAG* and *OitaSTK* are expressed in the column and ovary, particularly after pollination. At late developmental stages, *OitaAG* is weakly expressed in the inner tepals, while *OitaSTK* is expressed at very low levels during late developmental stages in lip and root tissues. Notably, their expression pattern partially resembles the behavior of *Phalaenopsis* genes *PeMADS1* (C-class) and

PeMADS7 (D-class), which are first tenuously expressed in the immature ovary, and then show an increase in expression after pollination (Salemme et al. 2013).

7.4.4 E-Class Genes

In plants, MADS-box genes from the E-class are able to interact and influence the activity of other genes that determine the identity of floral organ during reproductive development (Teo et al. 2019). For instance, the SEPALLATA (SEP) proteins may form high-order oligomers with gene products encoded by genes from the A, B, and C-classes (Puranik et al. 2014). The most notable of these genes are the *Arabidopsis*loci *SEP3* (*AGL9*) and *SEP1*, *SEP2*, *SEP4* (*AGL2/3/4*) (Zahn et al. 2006). Their developmental importance is evidenced by the aberrant phenotype of flowers from the triple mutant *sep1/sep2/sep3*, which consist entirely of sepal-like organs (Ditta et al. 2004; Pelaz et al. 2000). In rice, E-class genes are grouped into two clades: the *SEP*-clade (*OsMADS7*, *OsMADS45*, *OsMADS8*, *OsMADS24*) and the *LOFSEP*-clade (*OsMADS1*, *OsMADS5*, *OsMADS34*) (Cui et al. 2010; Gao et al. 2010). The simultaneous knockdown of *SEP*- and *LOFSEP*-clade genes *OsMADS1*, *OsMADS5*, *OsMADS7* and *OsMADS8* induces a homeotic shift that turns all floral organs (except the lemma) into leaf-like tissues. Reduction and loss of activity in *OsMADS34* activity lead to changes in panicle morphology, including alterations in the number of primary and secondary branches and a reduction in spikelet formation. Moreover, the *osmads34/osmads1* mutant shows defects in the specification of organ identity in the lemma/palea, in lodicules, in stamens, and in carpels. These phenotypes constitute evidence that support the hypothesis that *OsMADS34* plays a crucial role during the determination of development in panicles and spikelets (Gao et al. 2010). Mutation of the petunia *FBP2* gene, a presumed ortholog of *SEP3*, causes development of greenish petals and growth of ectopic inflorescences in the third floral whorl, while mutation of *FBP5* (a *SEP1/2* ortholog) does not lead to any obvious flower phenotypes. Surprisingly, the *fbp2/fbp5* double mutant shows developmental conversion of floral organs into leaf-like structures (Vandenbussche et al. 2003b). Interpretation of these results strongly suggests that the family of *SEP* genes may play indispensable roles during the formation of petal, stamen, and carpel formation in higher plants. Luckily many *SEP-like* genes have been characterized at the molecular level in a few orchid species (Lu et al. 1993). The *OM1* locus was the first E-class gene to be identified from tissues of mature flowers of *Aranda* Deborah and is expressed very weakly in petals and sepals, but not in the column (Lu et al. 1993). Genes *DOMADS1* and *DOMADS3* from *Dendrobium* Madame Thong-In are the presumed homologous of *Arabidopsis SEP3* and *SEP4* (Yu and Goh 2000), while the *DcOSEP1* locus from *D. crumenatum* is the putative homolog of *Arabidopsis SEP3* (Xu et al. 2006). In the case of *DOMADS1* and *DOMADS3*, both are expressed in the transitional shoot apical meristem during the floral transition and later on in tissues of mature flowers (Yu and Goh 2000). The *DOMADS1* transcript is consistently expressed first in the nascent inflorescence meristems and floral primordia, and then across all floral organs. Quite interestingly, the expression profile of *DOMADS1* that is observed in mature flowers is almost identical to that of *DcOSEP1* in *D. crumenatum* and *SEP1* in *Arabidopsis thaliana* (Pelaz et al. 2000; Xu et al. 2006; Yu and Goh 2000). The initiation of active transcription occurs in the early shoot apical meristem just prior the differentiation of the very first flower primordium (Yu and Goh 2000), while the *DOMADS3* transcript is barely detected in tissues of the pedicel. Thus, the *DOMADS3* gene represents a very important regulatory factor during the early floral transition and during the development of the pedicel (Yu and Goh 2000).

In mature flowers of *Oncidium,* the *OMADS6* transcript is highly expressed in tissues such as sepals, petals, the lip, the carpel, the anther cap, the stigmatic cavity and in the column, albeit weakly. The *OMADS11* gene is believed to be a

*SEP1/2*homolog that displays an expression pattern reminiscent to that of *OMADS6* (Chang et al. 2009; Hsu et al. 2003). Overexpression in *Arabidopsis* of the coding sequences for *OMADS6* and 11 induces very precocious flowering. Also, floral organ conversion is observed in plants transformed with the *35S::OMADS6* construct, including development of carpelloid sepals and staminoid petals, but this phenotype is not observed in plants transformed with the *35S:: OMADS10* construct, suggesting perhaps that *OMADS6* operates upstream of *OMADS10* (Chang et al. 2009).

In the emerging orchid model, *P. equestris*, four *SEP-like* have been cloned and characterized, which allowed them to be phylogenetically grouped into two clades: *PeSEP1/3* and *PeSEP2/4* (Pan et al. 2014). Apparently, these *PeSEP* genes are ubiquitously expressed across all floral organs. Results from yeast two-hybrid assays indicate that these PeSEP proteins may form high-order complexes with other suspected MADS-box proteins such as those from the B, C, D classes and with a putative AGAMOUS 6 homolog, and that these interactions may be instrumental during the determination of organ identity in flowers (Pan et al. 2014). Virus-induced gene silencing (VIGS) of *PeSEP3* results in development of leaf-like tepals, with visible alterations in epidermal cell identity and with changes in the cytoplasmic content of anthocyanin and chlorophyll. While the silencing of *PeSEP2* causes a mild defective phenotype, simultaneous silencing of both *PeSEP2* and *PeSEP3* reduces the expression of other B-class genes such as *PeMADS2-6*. This molecular phenotype suggests a link between *PeSEP* function and the expression of genes from the B-class. This result might also be interpreted as revealing functional diversification of the *Phalaenopsis PeSEP* family by forming multiple protein complexes (Pan et al. 2014).

In *C. goeringii*, four prospective E-class genes named *CgSEP1-4* have been functionally characterized in the wild type and in a peloric mutant (Xiang et al. 2018). The expression of *CgSEP1* is the highest in the lip of the wild type but the lowest in the sepal of the peloric mutant. In the case of *CgSEP3*, expression peaks in the peloric mutant within the lip-like petals and in the lip, while the opposite pattern is observed for *CgSEP4*. Finally, the expression of *CgSEP2* seems to reach a peak in the sepal, and sinks to a relative bottom in the lip and column of both the wild type and the mutant, perhaps suggesting it behaves as an enabling factor to other genes. The expression of other two putative *SEP*-like genes (*HrSEP-1* and *HrSEP-2*) was studied in the wild type and in a peloric cultivar of *Habenaria radiata* called 'Ryokusei' (Mitoma and Kanno 2018). The *HrSEP-1* and *HrSEP-2* transcripts are expressed in all the organs of wild-type flowers. However, only *HrSEP-2* is transcribed in the mutant, possibly because of the insertion of transposons within the coding sequence of *HrSEP-1*. Results from yeast two-hybrid assays suggest that HrSEP-1 interacts stably with HrDEF, HrAG-1, and HrAG-2, lending further experimental evidence to the hypothesis that loss of function at the *HrSEP-1* locus causes the mutant flower phenotype seen in 'Ryokusei'. In summary, the activity of MADS-box gene *HrSEP-1* is believed to be a key determinant in the development of petals, the column and the lip of *Habenaria radiata* (Mitoma and Kanno 2018).

7.4.5 *AGL6-like* Genes

AGL6-like genes are one of the most numerous clades within the family of MADS-box transcription factors. They are phylogenetically related to the *AGL2* (E-class) and *SQUA* (A-class) gene subfamilies (Ma et al. 1991). The *Arabidopsis* genes *AGL6* and *AGL13* are key representatives of the *AGL6* subfamily (Ma et al. 1991; Rounsley et al. 1995). *AGL6* gene is believed to control a wide range of processes such as the development of lateral organs, the initiation of flowering (Koo et al. 2010), the circadian clock (Yoo et al. 2011a), exert negative regulation on the *FLC/MAF* (*FLOWERING LOCUS C/MADS AFFECTING* FLOWERING) clade of genes, and perform positive regulation of *FT* (*FLOWERING LOCUS T*) (Yoo et al. 2011b). The amino acid sequence of putative

AGL6-like proteins shows: a) several amino acid substitutions in highly conserved residues of the amino-terminal MADS-box domain, and: b) the sequence of motifs I and II in the variable carboxy-terminus is poorly conserved (Dirks-Mulder et al. 2017). In monocots, the AGL6 family comprises four identifiable clades, and most orchid homologs appear to belong to clades 3 and 4 (Dirks-Mulder et al. 2017).

Only one candidate *AGL6* homolog gene has been reported in petunia, the *PETUNIA MADS BOX GENE4* (*pMADS4* or *PhAGL6*) (Tsuchimoto et al. 2000). Strong expression of *PhAGL6* is observed at the petal primordia, the early ovary primordium, the late ovule primordia, while expression is low in young developing anthers and barely detectable in sepals and anthers (Tsuchimoto et al. 2000). The null mutant *phagl6* shows no morphological abnormalities in none of the flower organs or in the ovules, the average seedset is not reduced and there are no changes in the flowering time (Tsuchimoto et al. 2000). However, in the *phagl6/fbp2* double mutant the petals are totally green and the corolla looks significantly reduced in size (Tsuchimoto et al. 2000). The stamens of the double mutant often develop into sepal or petal-like organs, and the top of the anthers turns into stigma-like structures. Yeast two-hybrid experiments suggest that PhAGL6 and FBP2 interact with proteins from the C class and perhaps with other SEP proteins, indicating that perhaps PhAGL6 behaves as an E-class protein in tissues of the fourth whorl (Tsuchimoto et al. 2000).

The rice genome contains two putative *AGL6-like* genes, *OsMADS6* and *OsMADS17*. The expression of *OsMADS6* takes place in the early floral meristem, and later in the primordium of the emerging palea (Li et al. 2010; Ohmori et al. 2009). The presumed *osmads6* mutant shows alterations in the development of the palea and lodicule and grows extra carpels and spikelets, while the *osmads6/osmads17* (*mfo1/lhs1*) double mutant showed more severe defects such as loss of determinacy at the spikelet meristem, extra abnormal spikelets without internal flower organs, extra defective glume-like organs without either any inner floral organs or with abnormal lemma and palea. These results were interpreted as suggesting that *OsMADS6* may specify a basal developmental floral state while *OsMADS17* defines identity at late stages. This result may suggest that *OsMADS6* is functionally similar to other E-class genes, which are expected to regulate the development of all four floral whorls and control determinacy at the floral meristem (Li et al. 2010). On the other hand, expression of *OsMADS17* is only observed in very young lodicules promordia, but there is no subsequent detection in palea primordia (Reinheimer and Kellogg 2009). The grass-relative *Joinvillea ascendens* develops its flowers like a typical monocot, in which the sterile organs are placed in the inner and outer tepals, which correspond to different whorls (Reinheimer and Kellogg 2009). The *JaAGL6* transcript is first observed at the primordia of the inner tepals, in the young anthers, in the gynoecium and relatively less in the inner tepals and in the carpel (Reinheimer and Kellogg 2009). By comparing the expression pattern of *AGL6-likes* across *Joinvillea*, *Oryza sativa*, *Streptochaeta angustifolia*, *Eleusine indica*, *Lolium temulentum*, *Triticum monococcum*, *Setaria italica* and *Sorghum bicolor*, it was possible to observe a pattern: expression consistently occurs in ovules, lodicules, the palea, and floral meristems, while expression is not significant in leaves, the culm, and in the root. Of outstanding reproductive relevance is the finding that expression of *AGL6-like* genes in tissues of the palea is absolutely retained in all spikelet-bearing species. This fact implies that *AGL6-like* genes play a conserved, decisive role during development of the palea (Reinheimer and Kellogg 2009).

The *Oncidium* genome features two putative orchid *AGL6-like* genes (*OMADS7/1*) (Chang et al. 2009). The *OMADS7* transcript is noticeable in flower buds at different developmental stages, in the sepals, petals, lip, carpel, anther cap and stigmatic cavity, but it is relatively harder to detect in the stamen. In fact the expression pattern of *OMADS7* resembles what has been reported for the E-class gene *OMADS6* and for other *AGL6-like* genes such as founding member

AGL6 in *Arabidopsis thaliana* and *ZAG3* in maize (Chang et al. 2009). Ectopic overexpression of the *OMADS7* sequence in *Arabidopsis* causes a phenotype to that is similar to the phenotype induced by the *35S::OMADS6* construct in other plants, including development of small curled leaves, early flowering and conversion of sepals into carpel-like structures (Chang et al. 2009). The *OMADS1* transcript is present in the apical meristem, in the lip and in the carpel. Biochemical characterization by the yeast two-hybrid assay indicates that OMADS1 can interact strongly with OMADS3. Ectopic expression of *OMADS1* caused a reduction in plant size, extremely early flowering, and loss of inflorescence determinacy (Hsu et al. 2003). Other phenotypes observed are homeotic conversion of sepals into carpel-like organs and of petals into aberrant staminoid organs (Hsu et al. 2003). In the very commercially important *Phalaenopsis* species *P. aphrodite*, an *AGL6-like* candidate gene named *PaAGL6-1* is expressed specifically in the lip, suggesting that *PaAGL6* may indeed play a prominent role during the formation of the lip (Su et al. 2013b).

In *Phalaenopsis* Brother Spring Dancer 'KHM190', the putative genes *PhAGL6a* and *PhAGL6b* are highly expressed in the lip of the wild type, while in a peloric mutant they appear to be differently upregulated in lip-like petals and lip-like sepals (Huang et al. 2016). Notably the expression of the *PhAGL6b* transcript is significantly reduced in a mutant with large lips, while no change is observed for *PhAGL6a*. In the mutant with large lips, the labellum progressively develops into a large petal-like which did not occur in the wild-type flower. To better understand transcriptional dynamics in the mutant, the full-length sequence of *PhAGL6b* was cloned from cDNA libraries harvested from lip tissues of the wild-type, a peloric mutant and the big lip mutant. Unexpectedly, four *PhAGL6b* transcripts showing alternatively splicing were identified from petaloid tissue of the big lip mutant (Fig. 7.2a, b).

To determine if the alternatively spliced isoforms of *PhAGL6b* influence the conversion of the lip into a petal-like organ in the so-called big lip mutant, RT-PCR was performed on total RNA extracted from the lip of flowers from the wild type and from the mutant. In tissues from the mutant, it was possible to detect 500 $\sim$ 700 bp bands possibly corresponding to alternatively spliced *PhAGL6b* transcripts, which had not been reported in orchids (Huang et al. 2016). Apparently the alternative splicing of *PhAGL6b* leads to the production of three different in-frame transcripts called *PhAGL6b-1*, *PhAGL6b-2* and *PhAGL6b-4*, whereas one frameshift transcript (*PhAGL6b-3*) is found only in the big lip mutant. Further studies using real-time PCR reveal that expression of the native non-spliced *PhAGL6b* is reduced by up to 70% in the big lip mutants, whereas all of the alternatively spliced isoforms show enhanced expression in the mutant. Taken as a whole, results from RT-PCR and real-time PCR experiments corroborated the hypothesis that specific expression of the alternatively spliced isoforms of *PhAGL6b* takes place in the petal-like lip of the big lip mutant. Thus, *PhAGL6b* is a truly determinant factor during the development of the lip in *Phalaenopsis*. The four *PhAGL6b* isoforms differ only in the length of the predicted C-domain (Huang et al. 2016). The C-domain is believed to be required for proper activation of transcription in target genes (Honma and Goto 2001) and may affect the degree of interaction with other MADS-box proteins in complexes (Gramzow and Theißen 2010; Geuten et al. 2006). Remarkably, the *Oncidium* L (lip) complex (OAP3-2/OAGL6-2/OAGL6-2/OPI) is believed to be required for lip formation (Hsu et al. 2015), and *Phalaenopsis PhAGL6b* happens to be a close homolog of *OAGL6-2*. Therefore it might be deduced that the *PhAGL6b* transcript and its splicing isoforms may compete with each other in its interaction with a hypothetical *Phalaenopsis* L-like complex during development of the labellum, as reported in *Oncidium* (Hsu et al. 2015). This hypothesis supports the notion that

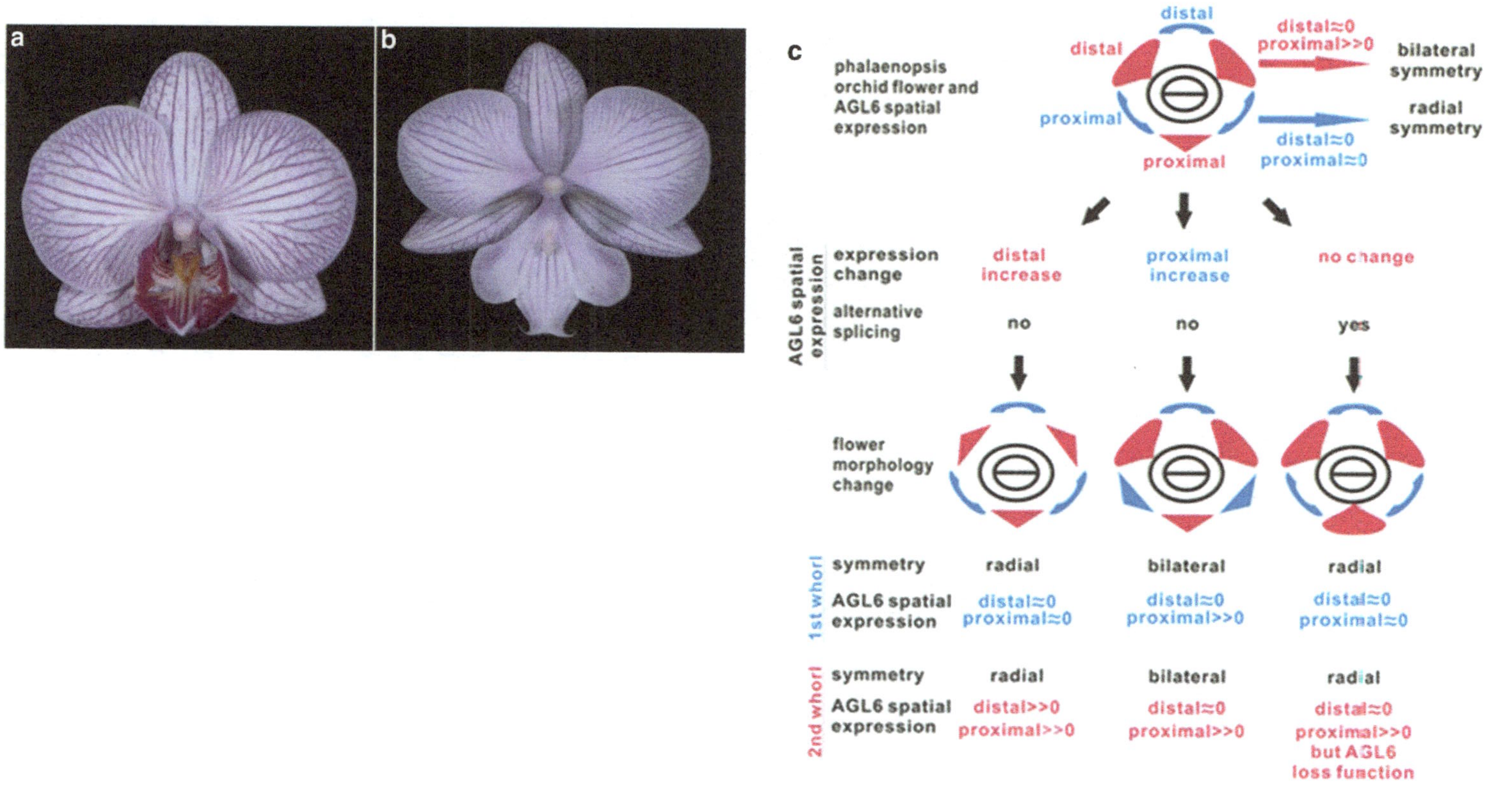

Fig. 7.2 Potential evolutionary relationship of *PhAGL6b* in the regulation of lip formation and floral symmetry in *Phalaenopsis* orchid. **a** Wild-type flower with zygomorphic symmetry. **b** A big lip mutant of *Phalaenopsis* from a breeding population. **c** Model of *PhAGL6b* spatial expression regulating *Phalaenopsis* floral symmetry. Ectopic expression of *PhAGL6b* in the distal domain (petal; pink) led to peloric petal conversion and change to radial symmetry. Ectopic expression in proximal domain, (sepal; blue) converts sepal to a lip-like peloric structure with bilateral symmetry. The alternative splicing of *PhAGL6b* transcripts produced in proximal domain (labellum; pink), converting the labellum to petal-like leads to radial symmetry. *PhAGL6b* expression patterns in *Phalaenopsis* floral organs are either an expansion or a reduction across labellum. To sum up, *PhAGL6b* may be a key regulator for floral symmetry evolvement. Pink color: 2nd whorl of the flower; blue color: 1st whorl of the flower; fan-shaped symbol: petal or petal-like structure; triangle symbol: labellum or lip-like structure; Curved symbol: sepal (Partially modified after Huang et al. 2016)

PhAGL6b may function as the most prominent floral regulator in *Phalaenopsis*, with wide impact on the determination of organ identity in petals, sepals and lip (Fig. 7.2c).

7.5 Conclusions and Perspectives

Orchid flowers are appreciated due to the wide array of colors, shapes and sensuous fragrances. The floral diversity of the Orchidaceae has long attracted the interest of evolutionary biologists. Among the most recently developed molecular methods, proteomics and genomics may be of unmatched value for the elucidation of the complex and diverse mechanisms that determine the organ development in orchid flowers. MADS-box proteins have been shown convincingly to play critical developmental roles during flowering and during floral patterning in orchid genera of commercial importance such as *Cymbidium*, *Oncidium*, *Dendrobium* and *Phalaenopsis*. A deep understanding of the molecular basis of orchid flowering and flower organ development may be applied to conventional orchid breeding and targeted locus manipulation to obtain novel flowering traits and flashy floral patterns. Orchid genomes may harbor many MADS-box genes, probably as many as 51 in *P. equestris* and 63 in *D. catenatum* (Cai et al. 2015; Zhang et al. 2016). Only a few of them have been properly characterized, but most have been shown to be involved one way or another in orchid flowering or floral organ development. For example, silencing of *OAGL6-2*, *SEP-like*, *PeMADS1*, *PeMADS7* by VIGS has been tested in *Oncidium* and *Phalaenopsis* hybrids to test new hypotheses (Hsieh et al. 2013; Hsu et al. 2015; Pan et al. 2014). To truly improve our understanding of the function of these orchid genes, more efficient and more stable genetic transformation methods are badly required for breeding of desired traits. Techniques related to the use of Clustered Regularly Interspaced Short Palindromic Repeats (CRISPR) and the CRISPR-associated gene Cas9 represent a step forward for the precise editing of genes in non-model organisms, which may prove useful for breeding and plant research. Recently, CRISPR-Cas9 has been successfully used to create multiple stable mutations in MADS-box genes (*PeMADS8*, *36*, *44*) of *P. equestris* (Tong et al. 2020). The draft genome has been published in *Apostasia shenzhenica*, *D. catenatum*, *D. officinale*, *P. equestris*, *P.* Brother Spring Dancer 'KHM190' and *Vanilla planifolia* (Yan et al. 2015; Zhang et al. 2017; Hu et al. 2019; Cai et al. 2015; Huang et al. 2016; Zhang et al. 2016). Moreover, the transcriptomes or the siRNAs of several orchid species are freely available through online databases such as Orchidstra 2.0 (http://orchidstra2.abrc.sinica.edu.tw), OrchidBase 3.0 (http://orchidbase.itps.ncku.edu.tw), OOGB (http://predictor.nchu.edu.tw/oogb) and PhalDB (http://www.yourgenebio.com/cylee2014/index) (Chang et al. 2011; Fu et al. 2011; Su et al. 2013a; Tsai et al. 2013; Chao et al. 2017; Tsai et al. 2017; Lee et al. 2018). Precise orchid production and marketing of superior potted varieties or cut flowers with specific novel traits may be within reach.

References

Aceto S, Gaudio L (2011) The MADS and the beauty: genes involved in the development of orchid flowers. Curr Genom 12:342–356

Acri-Nunes-Miranda R, Mondragón-Palomino M (2014) Expression of paralogous SEP-, FUL-, AG- and STK-like MADS-box genes in wild-type and peloric *Phalaenopsis* flowers. Front Plant Sci 5:76

Alvarez-Buylla ER, Pelaz S, Liljegren SJ, Gold S, Burgeff C, Ditta GS, Ribas de Pouplana L, Martinez-Castilla L, Yanofsky MF (2000) An ancestral MADS-box gene duplication occurred before the divergence of plants and animals. Proc Natl Acad Sci USA 10:5328–5333

Ambrose BA, Lerner DR, Ciceri P, Padilla CM, Yanofsky MF, Schmidt RJ (2000) Molecular and genetic analyses of the *silky1* gene reveal conservation in floral organ specification between eudicots and monocots. Mol Cell 5:569–579

Angenent GC, Colombo L (1996) Molecular control of ovule development. Trends Plant Sci 1:228–232

Angenent GC, Franken J, Busscher M, van Dijken A, van Went JL, Dons HJ, van Tunen AJ (1995) A novel class of MADS box genes is involved in ovule development in petunia. Plant Cell 7:1569–1582

Becker A, Theißen G (2003) The major clades of MADS-box genes and their role in the development and

evolution of flowering plants. Mol Phylogen Evol 29:464–489
Bowman JL, Drews GN, Meyerowitz EM (1991) Expression of the *Arabidopsis* floral homeotic gene AGAMOUS is restricted to specific cell types late in flower development. Plant Cell 3:749–758
Cai J, Liu X, Vanneste K, Proost S, Tsai WC, Liu KW, Chen LJ, He Y, Xu Q, Bian C, Zheng Z, Sun F, Liu W, Hsiao YY, Pan ZJ, Hsu CC, Yang YP, Hsu YC, Chuang YC, Dievart A, Dufayard JF, Xu X, Wang JY, Wang J, Xiao XJ, Zhao XM, Du R, Zhang GQ, Wang M, Su YY, Xie GC, Liu GH, Li LQ, Huang LQ, Luo YB, Chen HH, Van de Peer Y, Liu ZJ (2015) The genome sequence of the orchid *Phalaenopsis equestris*. Nat Genet 47:65–72
Chang YY, Chiu YF, Wu JW, Yang CH (2009) Four Orchid (*Oncidium* Gower Ramsey) *AP1/AGL9-like* MADS box genes show novel expression patterns and cause different effects on floral transition and formation in *Arabidopsis thaliana*. Plant Cell Physiol 50:1425–1438
Chang YY, Chu YW, Chen CW, Leu WM, Hsu HF, Yang CH (2011) Characterization of *Oncidium* Gower Ramsey transcriptomes using 454 GS-FLX pyrosequencing and their application to the identification of genes associated with flowering time. Plant Cell Physiol 52:1532–1545
Chang YY, Kao NH, Li JY, Hsu WH, Liang YL, Wu JW, Yang CH (2010) Characterization of the possible roles for B class MADS box genes in regulation of perianth formation in orchid. Plant Physiol 152:837–853
Chao YT, Yen SH, Yeh JH, Chen WC, Shih MC (2017) Orchidstra 2.0- A transcriptomics resource for the orchid family. Plant Cell Physiol 58:e9–e9
Chen D, Guo B, Hexige S, Zhang T, Shen D, Ming F (2007) *SQUA-like* genes in the orchid *Phalaenopsis* are expressed in both vegetative and reproductive tissues. Planta 226:369–380
Chen YY, Lee PF, Hsiao YY, Wu WL, Pan ZJ, Lee YI, Liu KW, Chen LJ, Liu ZJ, Tsai WC (2012) C-and D-class MADS-box genes from *Phalaenopsis equestris* (Orchidaceae) display functions in gynostemium and ovule development. Plant Cell Physiol 53:1053–1067
Christenson EA (2001) Phalaenopsis-a Monograph. Timber Press, Portland Oregon
Colombo L, Franken J, Koetje E, van Went J, Dons HJ, Angenent GC, van Tunen AJ (1995) The petunia MADS box gene *FBP11* determines ovule identity. Plant Cell 7:1859–1868
Colombo L, Franken J, Van der Krol AR, Wittich PE, Dons HJ, Angenent GC (1997) Downregulation of ovule-specific MADS box genes from petunia results in maternally controlled defects in seed development. Plant Cell 9:703–715
Cui R, Han J, Zhao S, Su K, Wu F, Du X, Xu Q, Chong K, Theißen G, Meng Z (2010) Functional conservation and diversification of class E floral homeotic genes in rice (*Oryza sativa*). Plant J 61:767–781
Davies B, Motte P, Keck E, Saedler H, Sommer H, Schwarz-Sommer Z (1999) *PLENA* and *FARINELLI*: redundancy and regulatory interactions between two *Antirrhinum* MADS-box factors controlling flower development. EMBO J 18:4023–4034
Dirks-Mulder A, Butôt R, van Schaik P, Wijnands JWPM, van den Berg R, Krol L, Doebar S, van Kooperen K, de Boer H, Kramer EM, Smets EF, Vos RA, Vrijdaghs A, Gravendeel B (2017) Exploring the evolutionary origin of floral organs of *Erycina pusilla*, an emerging orchid model system. BMC Evol Biol 17:89
Ditta G, Pinyopich A, Robles P, Pelaz S, Yanofsky MF (2004) The SEP4 gene of *Arabidopsis thaliana* functions in floral organ and meristem identity. Curr Biol 14:1935–1940
Dornelas MC, Patreze CM, Angenent GC, Immink RGH (2011) MADS: the missing link between identity and growth? Trends Plant Sci 16:89–97
Fornara F, Marziani G, Mizzi L, Kater M, Colombo L (2003) MADS-box genes controlling flower development in rice. Plant Biol 5:16–22
Fornara F, Pařenicová L, Falasca G, Pelucchi N, Masiero S, Ciannamea S, Lopez-Dee Z, Altamura MM, Colombo L, Kater MM (2004) Functional characterization of *OsMADS18*, a member of the *AP1/SQUA* subfamily of MADS box genes. Plant Physiol 135:2207–2219
Fu CH, Chen YW, Hsiao YY, Pan ZJ, Liu ZJ, Huang YM, Tsai WC, Chen HH (2011) OrchidBase: a collection of sequences of the transcriptome derived from orchids. Plant Cell Physiol 52:238–243
Gao X, Liang W, Yin C, Ji S, Wang H, Su X, Guo C, Kong H, Xue H, Zhang D (2010) The *SEPALLATA-like* gene *OsMADS34* is required for rice inflorescence and spikelet development. Plant Physiol 153:728–740
Geuten K, Becker A, Kaufmann K, Caris P, Janssens S, Viaene T, Theißen G, Smets E (2006) Petaloidy and petal identity MADS-box genes in the balsaminoid genera *Impatiens* and *Marcgravia*. Plant J 47:501–518
Goto K, Meyerowitz EM (1994) Function and regulation of the *Arabidopsis* floral homeotic gene *PISTILLATA*. Genes Dev 8:1548–1560
Gramzow L, Theißen G (2010) A hitchhiker's guide to the MADS world of plants. Genome Biol 11:214
Griesbach RJ (2002) Development of *Phalaenopsis* orchids for the mass-market. In: Janick J W (ed) Trends in New Crops and New Uses. ASHS Press, Alexandria, VA, pp 458–465
Hayes TE, Sengupta P, Cochran BH (1988) The human c-fos serum response factor and the yeast factors *GRM/PRTF* have related DNA-binding specificities. Genes Dev 2:1713–1722
Honma T, Goto K (2001) Complexes of MADS-box proteins are sufficient to convert leaves into floral organs. Nature 409:525–529
Hsieh MH, Pan ZJ, Lai PH, Lu HC, Yeh HH, Hsu -C, Wu -L, Chung MC, Wang SS, Chen WH, Chen HH (2013) Virus-induced gene silencing unravels multiple transcription factors involved in floral growth and

development in *Phalaenopsis* orchids. J Exp Bot 64:3869–3884

Hsu CC, Wu PS, Chen TC, Yu CW, Tsai WC, Wu K, Wu WL, Chen WH, Chen HH (2014) Histone acetylation accompanied with promoter sequences displaying differential expression profiles of B-class MADS-box genes for *Phalaenopsis* floral morphogenesis. PLoS ONE 9:e106033

Hsu HF, Hsieh WP, Chen MK, Chang YY, Yang CH (2010) C/D class MADS box genes from two monocots, orchid (*Oncidium* Gower Ramsey) and lily (*Lilium longiflorum*), exhibit different effects on floral transition and formation in *Arabidopsis thaliana*. Plant Cell Physiol 51:1029–1045

Hsu HF, Hsu WH, Lee YI, Mao WT, Yang JY, Li JY, Yang CH (2015) Model for perianth formation in orchids. Nature Plants 1:15046

Hsu HF, Huang CH, Chou LT, Yang CH (2003) Ectopic expression of an orchid (*Oncidium* Gower Ramsey) *AGL6-like* gene promotes flowering by activating flowering time genes in *Arabidopsis thaliana*. Plant Cell Physiol 44:783–794

Hsu HF, Yang CH (2002) An orchid (*Oncidium* Gower Ramsey) AP3-like MADS gene regulates floral formation and initiation. Plant Cell Physiol 43:1198–1209

Hu Y, Resende MFR, Bombarely A, Brym M, Bassil E, Chambers AH (2019) Genomics-based diversity analysis of *Vanilla* species using a *Vanilla planifolia* draft genome and Genotyping-By-Sequencing. Sci Rep 9:3416

Huang JZ, Lin CP, Cheng TC, Chang BCH, Cheng SY, Chen YW, Lee CY, Chin SW, Chen FC (2015) A de novo floral transcriptome reveals clues into *Phalaenopsis* orchid flower development. PLoS ONE 10: e0123474

Huang JZ, Lin CP, Cheng TC, Huang YW, Tsai YJ, Cheng SY, Chen YW, Lee CP, Chung WC, Chang BCH, Chin SW, Lee CY, Chen FC (2016) The genome and transcriptome of *Phalaenopsis* yield insights into floral organ development and flowering regulation. PeerJ 4:e2017

Jack T, Brockman LL, Meyerowitz EM (1992) The homeotic gene *APETALA3* of *Arabidopsis thaliana* encodes a MADS box and is expressed in petals and stamens. Cell 68:683–697

Jofuku KD, den Boer BG, Van Montagu M, Okamuro JK (1994) Control of *Arabidopsis* flower and seed development by the homeotic gene *APETALA2*. Plant Cell 6:1211–1225

Kaufmann K, Melzer R, Theißen G (2005) MIKC-type MADS-domain proteins: structural modularity, protein interactions and network evolution in land plants. Gene 347:183–198

Kim SY, Yun PY, Fukuda T, Ochiai T, Yokoyama J, Kameya T, Kanno A (2007) Expression of a *DEFICIENS-like* gene correlates with the differentiation between sepal and petal in the orchid, *Habenaria radiata* (Orchidaceae). Plant Sci 172:319–326

Koo SC, Bracko O, Park MS, Schwab R, Chun HJ, Park KM, Seo JS, Grbic V, Balasubramanian S, Schmid M, Godard F, Yun DJ, Lee SY, Cho MJ, Weigel D, Kim MC (2010) Control of lateral organ development and flowering time by the *Arabidopsis thaliana* MADS-box Gene *AGAMOUS-LIKE6*. Plant J 62:807–816

Kramer EM, Jaramillo MA, Di Stilio VS (2004) Patterns of gene duplication and functional evolution during the diversification of the *AGAMOUS* subfamily of MADS box genes in angiosperms. Genetics 166:1011–1023

Krizek BA, Fletcher JC (2005) Molecular mechanisms of flower development: an armchair guide. Nat Rev Genet 6:688–698

Lee CY, Viswanath KK, Huang JZ, Lee CP, Lin CP, Cheng TC, Chang BC, Chung WC, Chen FC (2018) PhalDB:A comprehensive database for molecular mining of the *Phalaenopsis* genome, transcriptome and miRNome. Genet Mol Res 17: GMR18051

Lee S, Jeon JS, An K, Moon YH, Lee S, Chung YY, An G (2003) Alteration of floral organ identity in rice through ectopic expression of OsMADS16. Planta 217:904–911

Leitch IJ, Kahandawala I, Suda J, Hanson L, Ingrouille MJ, Chase MW, Fay MF (2009) Genome size diversity in orchids: consequences and evolution. Ann Bot 104:469–481

Li H, Liang W, Jia R, Yin C, Zong J, Kong H, Zhang D (2010) The *AGL6-like* gene *OsMADS6* regulates floral organ and meristem identities in rice. Cell Res 20:299–313

Lin CS, Hsu CT, Liao DC, Chang WJ, Chou ML, Huang YT, Chen JJW, Ko SS, Chan MT, Shih MC (2016) Transcriptome-wide analysis of the MADS-box gene family in the orchid *Erycina pusilla*. Plant Biotechnol J 14:284–298

Litt A, Irish VF (2003) Duplication and diversification in the *APETALA1/FRUITFULL* floral homeotic gene lineage: implications for the evolution of floral development. Genetics 165:821–833

Lu ZX, Wu M, Loh CS, Yeong CY, Goh CJ (1993) Nucleotide sequence of a flower-specific MADS box cDNA clone from orchid. Plant Mol Biol 23:901–904

Münster T, Pahnke J, Di Rosa A, Kim JT, Martin W, Saedler H, Theißen G (1997) Floral homeotic genes were recruited from homologous MADS-box genes preexisting in the common ancestor of ferns and seed plants. Proc Natl Acad Sci USA 94:2415–2420

Ma H, Yanofsky MF, Meyerowitz EM (1991) *AGL1-AGL6*, an *Arabidopsis* gene family with similarity to floral homeotic and transcription factor genes. Genes Dev 5:484–495

Mandel MA, Yanofsky MF (1995) A gene triggering flower formation in *Arabidopsis*. Nature 377:522–524

Mao WT, Hsu HF, Hsu WH, Li JY, Lee YI, Yang CH (2015) The C-terminal sequence and PI motif of the orchid (*Oncidium* Gower Ramsey) *PISTILLATA* (*PI*) ortholog determine its ability to bind *AP3* orthologs and enter the nucleus to regulate downstream genes controlling petal and stamen formation. Plant Cell Physiol 56:2079–2099

Masiero S, Colombo L, Grini PE, Schnittger A, Kater MM (2011) The emerging importance of type I MADS box transcription factors for plant reproduction. Plant Cell 23:865–872

Messenguy F, Dubois E (2003) Role of MADS box proteins and their cofactors in combinatorial control of gene expression and cell development. Gene 316:1–21

Mitoma M, Kanno A (2018) The greenish flower phenotype of *Habenaria radiata* (Orchidaceae) is caused by a mutation in the *SEPALLATA-like* MADS-Box gene *HrSEP-1*. Front Plant Sci 9:831

Mondragón-Palomino M, Theißen G (2009) Why are orchid flowers so diverse? Reduction of evolutionary constraints by paralogues of class B floral homeotic genes. Ann Bot 13:51–59

Mondragón-Palomino M, Theißen G (2008) MADS about the evolution of orchid flowers. Trends Plant Sci 13:51–59

Mondragón-Palomino M, Theißen G (2011) Conserved differential expression of paralogous *DEFICIENS*- and *GLOBOSA-like* MADS-box genes in the flowers of Orchidaceae: refining the 'orchid code'. Plant J 66:1008–1019

Nagasawa N, Miyoshi M, Sano Y, Satoh H, Hirano H, Sakai H, Nagato Y (2003) *SUPERWOMAN1* and *DROOPING LEAF* genes control floral organ identity in rice. Development 130:705–718

Norman C, Runswick M, Pollock R, Treisman R (1988) Isolation and properties of cDNA clones encoding *SRF*, a transcription factor that binds to the *c-fos* serum response element. Cell 55:989–1003

O'Neill SD (1997) Pollination regulation of flower development. Annu Rev Plant Physiol Plant Mol Biol 48:547–574

Ohmori S, Kimizu M, Sugita M, Miyao A, Hirochika H, Uchida E, Nagato Y, Yoshida H (2009) *MOSAIC FLORAL ORGANS1*, an *AGL6-like* MADS box gene, regulates floral organ identity and meristem fate in rice. Plant Cell 21:3008–3025

Pan ZJ, Chen YY, Du JS, Chen YY, Chung MC, Tsai WC, Wang CN, Chen HH (2014) Flower development of *Phalaenopsis* orchid involves functionally divergent *SEPALLATA-like* genes. New Phytol 202:1024–1042

Pařenicová L, de Folter S, Kieffer M, Horner DS, Favalli C, Busscher J, Cook HE, Ingram RM, Kater MM, Davies B, Angenent GC, Colombo L (2003) Molecular and phylogenetic analyses of the complete MADS-box transcription factor family in *Arabidopsis*: new openings to the MADS world. Plant Cell 15:1538–1551

Passmore S, Maine GT, Elble R, Christ C, Tye BK (1988) *Saccharomyces cerevisiae* protein involved in plasmid maintenance is necessary for mating of MATα cells. J Mol Biol 204:593–606

Pelaz S, Ditta GS, Baumann E, Wisman E, Yanofsky MF (2000) B and C floral organ identity functions require *SEPALLATA* MADS-box genes. Nature 405:200–203

Petruzzello M (2018) List of plants in the family Orchidaceae. https://www.britannica.com/topic/list-of-plants-in-the-family-Orchidaceae-2075389

Pinyopich A, Ditta GS, Savidge B, Liljegren SJ, Baumann E, Wisman E, Yanofsky MF (2003) Assessing the redundancy of MADS-box genes during carpel and ovule development. Nature 424:85–88

Puranik S, Acajjaoui S, Conn S, Costa L, Conn V, Vial A, Marcellin R, Melzer R, Brown E, Hart D, Theißen G, Silva CS, Parcy F, Dumas R, Nanao M, Zubieta C (2014) Structural basis for the oligomerization of the MADS domain transcription factor SEPALLATA3 in *Arabidopsis*. Plant Cell 26:3603–3615

Purugganan MD, Rounsley SD, Schmidt RJ, Yanofsky MF (1995) Molecular evolution of flower development: diversification of the plant MADS-box regulatory gene family. Genetics 140:345–356

Rajkumari JD, Longjam RS (2005) Orchid flower evolution. J Genet 84:81–84

Reinheimer R, Kellogg EA (2009) Evolution of *AGL6-like* MADS box genes in grasses (Poaceae): ovule expression is ancient and palea expression is new. Plant Cell 21:2591–2605

Riechmann JL, Wang M, Meyerowitz EM (1996) DNA-binding properties of *Arabidopsis* MADS domain homeotic proteins *APETALA1*, *APETALA3*, *PISTILLATA* and *AGAMOUS*. Nucleic Acids Res 24:3134–3141

Rounsley SD, Ditta GS, Yanofsky MF (1995) Diverse roles for MADS box genes in *Arabidopsis* development. Plant Cell 7:1259–1269

Rudall PJ, Bateman RM (2002) Roles of synorganisation, zygomorphy and heterotopy in floral evolution: the gynostemium and labellum of orchids and other lilioid monocots. Biol Rev Camb Philos Soc 77:403–441

Salemme M, Sica M, Gaudio L, Aceto S (2013) The *OitaAG* and *OitaSTK* genes of the orchid *Orchis italica*: a comparative analysis with other C- and D-class MADS-box genes. Mol Biol Rep 40:3523–3535

Sawettalake N, Bunnag S, Wang Y, Shen L, Yu H (2017) *DOAP1* promotes flowering in the orchid *Dendrobium* Chao Praya Smile. Front Plant Sci 8:400

Schwarz-Sommer Z, Huijser P, Nacken W, Saedler H, Sommer H (1990) Genetic control of flower development by homeotic genes in *Antirrhinum majus*. Science 250:931–936

Skipper M, Johansen LB, Pedersen KB, Frederiksen S, Johansen BB (2006) Cloning and transcription analysis of an *AGAMOUS*- and *SEEDSTICK* ortholog in the orchid *Dendrobium thyrsiflorum* (Reichb. f.). Gene 366:266–274

Song IJ, Nakamura T, Fukuda T, Yokoyama J, Ito T, Ichikawa H, Horikawa Y, Kameya T, Kanno A (2006) Spatiotemporal expression of duplicate *AGAMOUS* orthologues during floral development in *Phalaenopsis*. Dev Genes Evol 216:301–313

Su CL, Chao YT, Yen SH, Chen CY, Chen WC, Chang YCA, Shih MC (2013a) Orchidstra: An integrated orchid functional genomics database. Plant Cell Physiol 54:e11–e11

Su CL, Chen WC, Lee AY, Chen CY, Chang YCA, Chao YT, Shih MC (2013b) A modified ABCDE model of flowering in orchids based on gene expression profiling studies of the moth orchid *Phalaenopsis aphrodite*. PLoS ONE 8:e80462

Swamy BGL (1943) Embryology of Orchidaceae. Curr Sci 12:13–17

Teo ZWN, Zhou W, Shen L (2019) Dissecting the function of MADS-box transcription factors in orchid reproductive development. Front Plant Sci 10:1474

Theißen G (2001) Development of floral organ identity: stories from the MADS house. Curr Opin Plant Biol 4:75–85

Theißen G, Becker A, Di Rosa A, Kanno A, Kim JT, Munster T, Winter KU, Saedler H (2000) A short history of MADS-box genes in plants. Plant Mol Biol 42:115–149

Theißen G, Melzer R (2007) Molecular mechanisms underlying origin and diversification of the angiosperm flower. Ann Bot 100:603–619

Theißen G, Melzer R, Rümpler F (2016) MADS-domain transcription factors and the floral quartet model of flower development: linking plant development and evolution. Development 143:3259–3327

Tian Y, Yuan X, Jiang S, Cui B, Su J (2013) Molecular cloning and spatiotemporal expression of an APETALA1/FRUITFULL-like MADS-box gene from the orchid (*Cymbidium faberi*). Sheng Wu Gong Cheng Xue Bao 29:203–213

Tong CG, Wu FH, Yuan YH, Chen YR, Lin CS (2020) High-efficiency CRISPR/Cas-based editing of *Phalaenopsis* orchid MADS genes. Plant Biotechnol J 18:889–891

Tsai WC, Dievart A, Hsu CC, Hsiao YY, Chiou SY, Huang H, Chen HH (2017) Post genomics era for orchid research. Bot Stud 58:61

Tsai WC, Fu CH, Hsiao YY, Huang YM, Chen LJ, Wang M, Liu ZJ, Chen H-H (2013) OrchidBase 2.0: comprehensive collection of Orchidaceae floral transcriptomes. Plant Cell Physiol 54:e7

Tsai WC, Lee PF, Chen HI, Hsiao YY, Wei WJ, Pan ZJ, Chuang MH, Kuoh CS, Chen WH, Chen HH (2005) *PeMADS6*, a *GLOBOSA/PISTILLATA-like* gene in *Phalaenopsis equestris* involved in petaloid formation, and correlated with flower longevity and ovary development. Plant Cell Physiol 46:1125–1139

Tsai WC, Pac ZJ, Hsiao YY, Chen LJ, Liu ZJ (2014) Evolution and function of MADS-box genes involved in orchid floral development. J Syst Evol 52:397–410

Tsai WC, Kuoh CS, Chuang MH, Chen WH, Chen HH (2004) Four *DEF-like* MADS box genes displayed distinct floral morphogenetic roles in *Phalaenopsis* orchid. Plant Cell Physiol 45:831–844

Tsuchimoto S, Mayama T, Van Der Krol A, Ohtsubo E (2000) The whorl-specific action of a petunia class B floral homeotic gene. Genes Cells 5:89–99

Tzeng TY, Yang CH (2001) A MADS box gene from lily (*Lilium longiflorum*) is sufficient to generate dominant negative mutation by interacting with *PISTILLATA* (*PI*) in *Arabidopsis thaliana*. Plant Cell Physiol 42:1156–1168

Valoroso MC, Censullo MC, Aceto S (2019) The MADS-box genes expressed in the inflorescence of *Orchis italica* (Orchidaceae). PLoS ONE 14:e0213185

Vandenbussche M, Theißen G, Van de Peer Y, Gerats T (2003a) Structural diversification and neo-functionalization during floral MADS-box gene evolution by C-terminal frameshift mutations. Nucleic Acids Res 31:4401–4409

Vandenbussche M, Zethof J, Souer E, Koes R, Tornielli GB, Pezzotti M, Ferrario S, Angenent GC, Gerats T (2003b) Toward the analysis of the *Petunia* MADS box gene family by reverse and forward transposon insertion mutagenesis approaches: B, C, and D floral organ identity functions require *SEPALLATA-like* MADS box genes in *Petunia*. Plant Cell 15:2680–2693

Wang SY, Lee PF, Lee YI, Hsiao YY, Chen YY, Pan ZJ, Liu ZJ, Tsai WC (2011) Duplicated C-class MADS-box genes reveal distinct roles in gynostemium development in *Cymbidium ensifolium* (Orchidaceae). Plant Cell Physiol 52:563–577

Weigel D, Meyerowitz EM (1994) The ABCs of floral homeotic genes. Cell 78:203–209

Whipple CJ, Ciceri P, Padilla CM, Ambrose BA, Bandong SL, Schmidt RJ (2004) Conservation of B-class floral homeotic gene function between maize and *Arabidopsis*. Development 131:6083–6091

Winter KU, Weiser C, Kaufmann K, Bohne A, Kirchner C, Kanno A, Saedler H, Theißen G (2002) Evolution of class B floral homeotic proteins: obligate heterodimerization originated from homodimerization. Mol Biol Evol 19:587–596

Wittich PF, de Heer RF, Cheng XF, Kieft H, Colombo L, Angenent GCand A. A. M. van Lammeren (1999) Immunolocalization of the petunia floral binding proteins 7 and 11 duringseed development in wild-type and expression mutants of *Petunia hybrida*. Protoplasma 208:224–229

Xiang L, Chen Y, Chen L, Fu X, Zhao K, Zhang J, Sun C (2018) B and E MADS-box genes determine the perianth formation in *Cymbidium goeringii* Rchb.f. Physiol Plant 162:353–369

Xu Y, Teo LL, Zhou J, Kumar PP, Yu H (2006) Floral organ identity genes in the orchid *Dendrobium crumenatum*. Plant J 46:54–68

Yan L, Wang X, Liu H, Tian Y, Lian J, Yang R, Hao S, Wang X, Yang S, Li Q, Qi S, Kui L, Okpekum M, Ma X, Zhang J, Ding Z, Zhang G, Wang W, Dong Y, Sheng J (2015) The genome of *Dendrobium officinale* illuminates the biology of the important traditional Chinese orchid herb. Mol Plant 8:922–934

Yang F, Zhu G (2015) Digital gene expression analysis based on de novo transcriptome assembly reveals new genes associated with floral organ differentiation of the orchid plant *Cymbidium ensifolium*. PLoS ONE 10: e0142434

Yang F, Zhu G, Wei Y, Gao J, Liang G, Peng L, Lu C, Jin J (2019) Low-temperature-induced changes in the transcriptome reveal a major role of *CgSVP* genes in regulating flowering of *Cymbidium goeringii*. BMC Genom 20:53

Yanofsky MF, Ma H, Bowman JL, Drews GN, Feldmann KA, Meyerowitz EM (1990) The protein encoded by the *Arabidopsis* homeotic gene agamous resembles transcription factors. Nature 346:35–39

Yoo SK, Hong SM, Lee JS, Ahn JH (2011a) A genetic screen for leaf movement mutants identifies a potential role for *AGAMOUS-LIKE 6* (*AGL6*) in circadian-clock control. Mol Cells 31:281–287

Yoo SK, Wu X, Lee JS, Ahn JH (2011b) *AGAMOUS-LIKE 6* is a floral promoter that negatively regulates the *FLC/MAF* clade genes and positively regulates *FT* in *Arabidopsis*. Plant J 65:62–76

Yu H, Goh CJ (2000) Identification and characterization of three orchid MADS-box genes of the *AP1/AGL9* subfamily during floral transition. Plant Physiol 123:1325–1336

Zahn LM, Leebens-Mack JH, Arrington JM, Hu Y, Landherr LL, DePamphilis CW, Becker A, Theißen G, Ma H (2006) Conservation and divergence in the *AGAMOUS* subfamily of MADS-box genes: evidence of independent sub- and neofunctionalization events. Evol Dev 8:30–45

Zhang GQ, Liu KW, Li Z, Lohaus R, Hsiao Y-Y, Niu SC, Wang JY, Lin YC, Xu Q, Chen LJ, Yoshida K, Fujiwara S, Wang ZW, Zhang YQ, Mitsuda N, Wang M, Liu GH, Pecoraro L, Huang HX, Xiao XJ, Lin M, Wu XY, Wu WL, Chen YY, Chang SB, Sakamoto S, Ohme-Takagi M, Yagi M, Zeng SJ, Shen CY, Yeh CM, Luo YB, Tsai WC, Van de Peer Y, Liu ZJ (2017) The *Apostasia* genome and the evolution of orchids. Nature 549:379–383

Zhang GQ, Xu Q, Bian C, Tsai WC, Yeh CM, Liu KW, Yoshida K, Zhang LS, Chang SB, Chen F, Shi Y, Su YY, Zhang YQ, Chen LJ, Yin Y, Lin M, Huang H, Deng H, Wang Z-W, Zhu SL, Zhao X, Deng C, Niu SC, Huang J, Wang M, Liu G-H, Yang HJ, Xiao XJ, Hsiao YY, Wu WL, Chen YY, Mitsuda N, Ohme-Takagi M, Luo YB, Van de Peer Y, Liu ZJ (2016) The *Dendrobium catenatum* Lindl. genome sequence provides insights into polysaccharide synthase, floral development, and adaptive evolution. Sci Rep 6:19029

8 Genomics, Transcriptomics and miRNA Family Resources for *Phalaenopsis aphrodite* and the Orchid Family

Ya-Ting Chao, Wan-Chieh Chen, Hsiu-Yin Ho, and Ming-Che Shih

Abstract

Phalaenopsis aphrodite, an epiphytic orchid native to Taiwan and Philippines, possesses many desirable horticultural characteristics permitting its use as parents in breeding commercial orchid hybrids. In order to facilitate the development of genomics/omics breeding tools for orchids, our effort was geared to making *P. aphrodite* gene annotations and transcriptome resources more accessible to the orchid research community. In this chapter, we review our major progress in *P. aphrodite* genomics and functional genomics, focusing on the genome sequencing and annotation, whole transcriptome analysis of flower development, analysis of orchid miRNA, and the integrative functional genomics database for the orchid family.

Y.-T. Chao · W.-C. Chen · H.-Y. Ho · M.-C. Shih (✉)
Agricultural Biotechnology Research Center, Academia Sinica, Taipei, Taiwan
e-mail: mcshih@gate.sinica.edu.tw

8.1 Introduction

The *Phalaenopsis* is in the subfamily Epidendroideae of Orchidaceae and its hybrids are popular potted ornamental plants. To date at least 70 *Phalaenopsis* species and over 34,000 hybrids are registered in the Royal Horticultural Society (Christenson 2001; Hsu et al. 2018).

P. aphrodite is mainly distributed in the southeastern Taiwan, and the northern Philippines. *P. aphrodite* carries many good flower traits, such as large flower size, round petals, overlapping floral segments, multiple flowers per inflorescence and many side branches of stalks (Christenson 2001). For this reason, *P. aphrodite* has been used as breeding parent to produce many superb commercial orchid hybrids, including the prestigious tetraploid hybrid *Phalaenopsis* Doris (*P.* Elisabethae × *P.* Katherine Siegwart), which was routinely used for breeding robust and colorful *Phalaenopsis* hybrids (Christenson 2001). According to the OrchidWiz database (https://www.orchidwiz.com/), more than 90% *Phalaenopsis* hybrid orchids are descendants of *P. aphrodite* (Hsu et al. 2018). In addition to its importance for orchid breeding, *P. aphrodite* has 19 pairs of chromosomes and a haploid genome size of 1.2 Gb, which is relatively small among the orchid species. Therefore, it was selected as our model orchid for functional genomics and genome research.

F.-C. Chen and S.-W. Chin (eds.), *The Orchid Genome*, Compendium of Plant Genomes,
https://doi.org/10.1007/978-3-030-66826-6_8

Since 2013, we have been developing the Orchidstra database (Su et al. 2013a). The current Orchidstra 2.0 database is an integrative functional genomics database for *P. aphrodite* and many other orchid species from 12 genera in five Orchidaceae subfamilies (Chao et al. 2017). Emphasis has been placed on accommodating the orchid transcriptome assembly/functional annotations, orthologous gene groups and processing of microarray/RNA-Seq gene expression data. In parallel with our orchid transcriptome project, we launched the *P. aphrodite* genome sequencing project with the aim of producing a high-quality reference genome. In addition to de novo genome assembly, we built a high-density genetic linkage map and a high-resolution pachytene karyotype for *P. aphrodite*. To the best of our knowledge, the *P. aphrodite* reference genome is the first orchid genome to be integrated with a genetic linkage map and cytogenetically validated by fluorescence in situ hybridization (FISH) (Chao et al. 2018).

8.2 *Phalaenopsis aphrodite* Genomic Resources

8.2.1 Chromosome-Level Assembly and Genome Annotation for *P. aphrodite*

The de novo assembly of *P. aphrodite* genome was performed using ALLPATHS-LG (Gnerre et al. 2011) with about 52 folds overlapped paired-end reads and 54 folds mate-pair reads from libraries with insert sizes ranging from 3 kb to 7 kb. To improve the scaffold continuity, the ALLPATHS-LG scaffolds were further linked using SSPACE (Boetzer et al. 2011) with mate-pair libraries ranging from 8 to 15 kb (a total coverage of 49X) and the 40 kb fosmid library (a coverage of 1.5X). Over 98% of the Orchidstra 2.0 full-length ESTs can be aligned over 90% of their length to a single scaffold (Chao et al. 2018).

In order to construct the genetic linkage map for orchid species, an F1 mapping population was produced by a cross between *P. aphrodite* and *P. modesta*. PstI-digested RADSeq (Restriction site Associated DNA Sequencing) was performed for the parents and their 184 progeny (Chao et al. 2018). We used the RAD-Seq data to identify SNPs and then constructed a high-density orchid genetic linkage map, which contains a total of 2,905 RAD markers covering 19 chromosomes. The linkage map has a total genetic length of 3075.8 cM, and the average interval between markers is about 1 cM. The scaffolds were anchored and oriented according to the RAD markers on the linkage groups (Fig. 8.1a). Several linkage group-specific genomic clones were constructed for each linkage group, and then these clones were localized on *P. aphrodite* pachytene chromosomes by FISH mapping. The FISH mapping results showed that all mapped linkage group-specific markers are in the same order as their positions on genetic map (Fig. 8.1b, c). FISH probes were also used to validate the within-scaffold arrangement and orientation. Based on the genetic linkage map and FISH mapping results, a chromosome-level genome assembly was achieved with a scaffold N50 of 19.7 Mb and a total length of 1025.1 Mb (Table 8.1). The completeness of the orchid reference genome was estimated to be 95% using BUSCO assessment (Simao et al. 2015).

8.2.2 Gene Content of the *P. aphrodite* Genome

The *P. aphrodite* genome contains 28,902 protein-coding genes (Fig. 8.2), of which 23,188 were supported by both RNA-Seq evidence and protein homology. Currently, 70.4% of the annotated genes have one or more InterPro hits, 88.1% of the genes have GO terms assigned, and 19.6% of the genes have obtained pathway annotations by searching KEGG (Table 8.1). Overall, 99.5% of the total protein-coding genes were assigned with at least one functional annotation.

The gene density is about 2.82 genes per 100 kb, and the average gene length (including

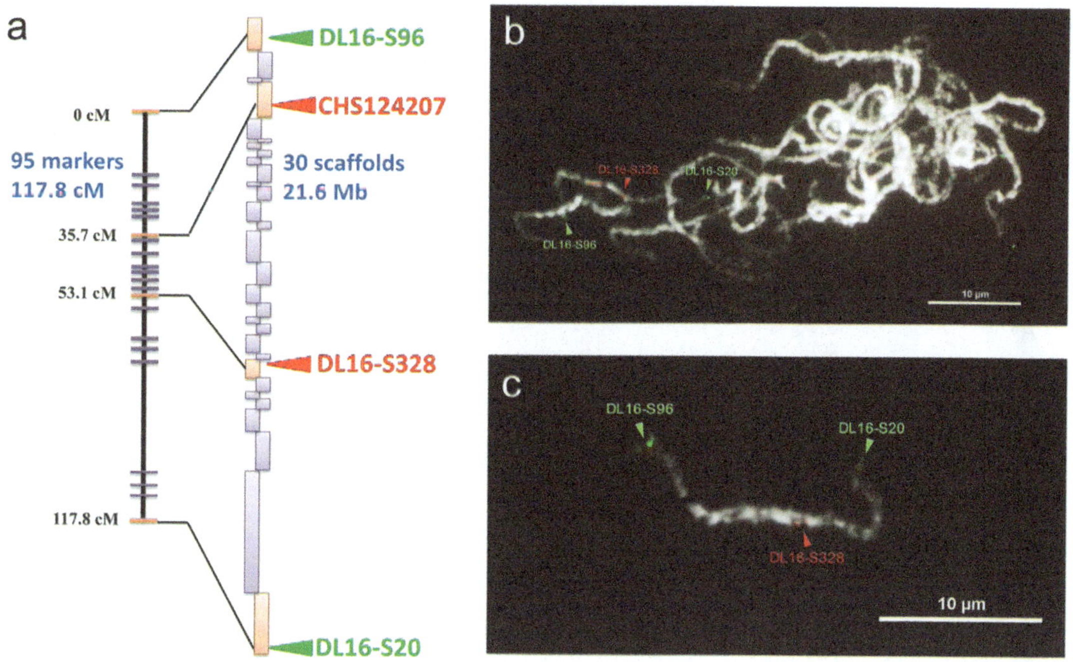

Fig. 8.1 Integration of genome assembly with genetic map and FISH mapping. Linkage group 19 is used as an example (Adapted from Chao et al. 2018)

Table 8.1 Genome assembly and protein-coding gene annotations in *P. aphrodite*. (Data from Chao et al. 2018)

Metrics for genome assembly	
Total scaffold length (Mb)	1025.1
Scaffold N50 (kb)	946.4
Scaffold N90 (kb)	69.4
Number of scaffolds > = 1 kb	13485
Max scaffold after integration with linkage map (Mb)	44.49 (Chromosome 2)
Scaffold N50 after integration with linkage map (Mb)	19.68
Number of protein-coding genes	28902
Average exon length (bp)	283
Average number of exons per gene	4.82
Average intron length (bp)	2665
Metrics for gene annotation	
Database	Number of genes annotated
NCBI nr	28738 (99.4%)
Swiss-Prot	16364 (56.6%)
InterPro	20357 (70.4%)
Gene Ontology by BLAST2GO	25467 (88.1%)
Enzyme Commission (EC) assignments by BLAST2GO KEGG mapping	5665 (19.6%)
No annotation	149 (0.5%)

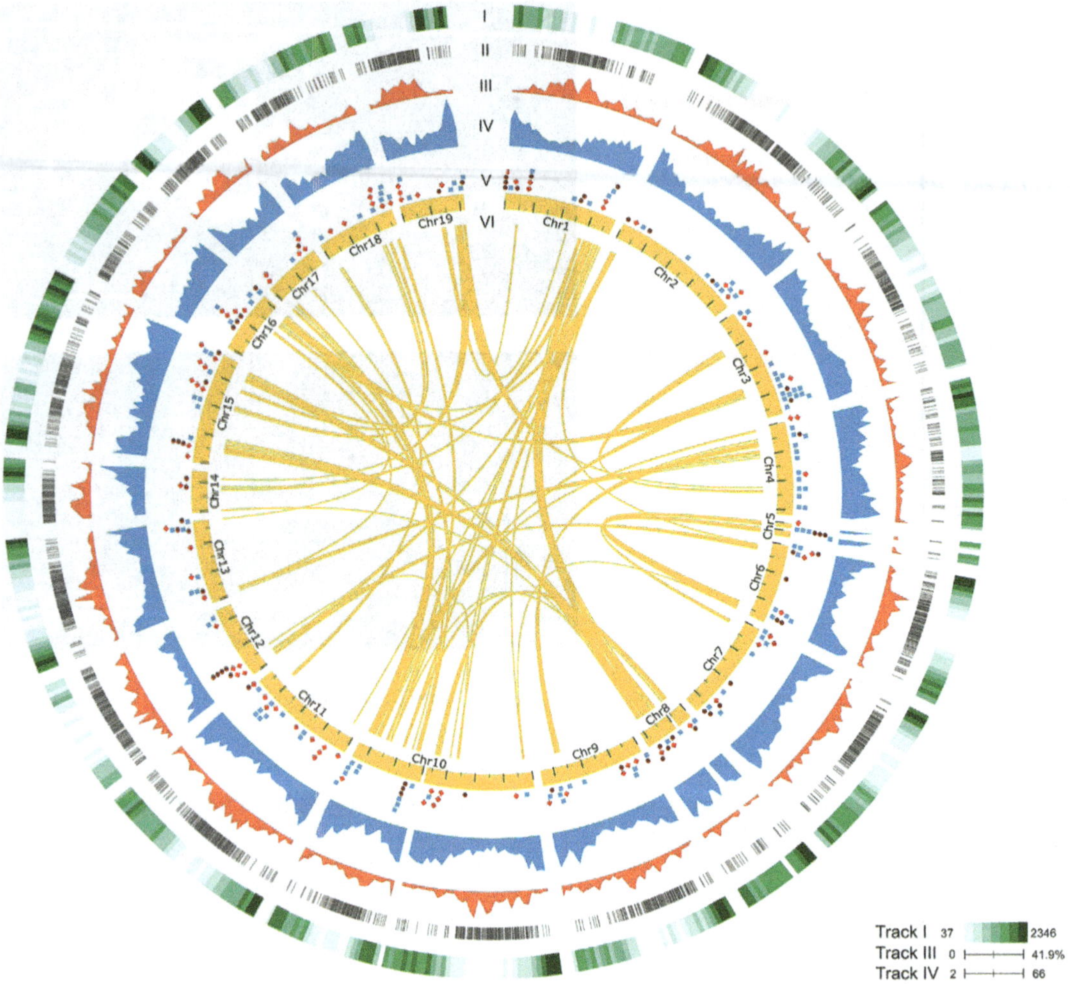

Fig. 8.2 Features of *P. aphrodite* genome, which is composed of 19 chromosomes (Adapted from Chao et al. 2018)

introns) is about 11.5 kb. In addition, 10,199 transposable element (TE)-related genes were identified. The protein-coding genes from *P. aphrodite*, rice, banana, tomato and Arabidopsis were used to identify a total of 19,912 orthologous groups, among which 8,612 groups were shared among the five species. In addition, 24,854 *P. aphrodite* genes were clustered into 12,084 orthologous/paralogous groups, including 690 *P. aphrodite*-specific groups.

8.2.3 FAR1/FRS Gene Family Associated with Adaptations to the Epiphytic Lifestyle

A total of 1,615 transcription factors were identified in the *P. aphrodite* genome, among which 216 loci belong to the FAR1 (far-red-impaired response 1)/FRS (FAR1-related sequence) gene family (Chao et al. 2018). The FAR1/FRS family contains six major clades, including the FRS3/5/9, FRS6/8, FRS10/11 clades that are composed of both dicot and monocot members, and the monocot-only FRS5-like clade. We found an orchid lineage-specific duplication event in the FRS 6/8 clade, and the FRS5-like clade contains an expanded subclade composed of 17 *P. aphrodite* genes and 14 genes from *P. equestris*/*D. catenatum*. In *A. thaliana*, the *FRS* genes are involved in the light signaling pathway controlling flowering time and development (Lin and Wang 2004). The orchid FRS3, FRS6, FRS10 genes are widely expressed in all of the tissues tested except pollinia, while the FRS11 are especially highly expressed in pollinia (Fig. 8.3). In *P. aphrodite*, the expression of FRS5-like genes is at lower levels but more tissue-specific, especially in flower buds (Fig. 8.3). It is suggested that the lineage-specific expansion of the orchid FRS5-like subclade might be a result of selective pressure exerted by unstable light of rainforest canopies. The tissue-specific expression of these FRS5 genes indicates that they may play roles in the regulation of light-sensing systems or flowering time in *P. aphrodite*.

8.2.4 The Anthocyanin Biosynthetic Genes in *P. aphrodite*

The *P. aphrodite* genome contains at least 347 genes associated with flowering time, floral color and scent. *P. aphrodite* is often used as breeding parent to produce large-flower orchid hybrids. The number of genes encoding flavonoid biosynthesis-related enzymes is similar for the white *P. aphrodite* and pink spotted *P. lueddemanniana*. In addition, the white *P. aphrodite* possesses orthologs of the known R2R3-MYB regulators of the anthocyanin biosynthetic pathway. RNA-Seq data of orthologous gene pairs were used to quantify gene expression in anthocyanin biosynthetic pathway in compatible tissue stage of *P. aphrodite* and *P. lueddemanniana*. For genes that are common to the synthesis of flavonoids and lignin, the expression patterns are similar in the two species. However, the genes involved in the anthocyanin branch pathways, including flavanone 3-hydroxylase (F3H), dihydroflavonol 4-reductase (DFR) and anthocyanidin synthase (ANS), exhibit differential expression patterns in *P. aphrodite* and *P. lueddemanniana* (Fig. 8.4). Both the genome annotation and differential gene expression analysis results suggest that the white flower trait was not resulted from the absence of known structural or regulatory genes. The *P. aphrodite* genomic sequence can provide a valuable resource for studying the regulatory elements controlling orchid floral trait.

8.3 Gene Expression Profiling on Orchid Floral Development

The model plant Arabidopsis flowers are radially symmetric, whereas orchid flowers are bilaterally symmetric with several unique features such as petalized sepals, a modified petal (lip) and coherent mass pollen grains (pollinia) (Fig. 8.5 a). Therefore, a classic ABC (or ABCDE) model of floral organ identity is not appropriated to directly apply to orchids (Mendoza et al. 1999). In our previous study, a modified model of flower tissue development based on gene expression patterns in *P. aphrodite* via Microarray technology was reported (Su et al. 2013b). MADS box genes are major players in this modified model. Analysis of the complete *P. aphrodite* genome sequence disclosed 56 genes encoding MADS box proteins, including 38 MIKC type and 18 M-type MADS genes (Table 8.2) (Chao et al. 2018).

Based on the comparison of gene expression patterns between Arabidopsis and *P. aphrodite*, orchid class A and B MADS box genes exhibited more diversified patterns than the genes in Arabidopsis (Fig. 8.5b). For example, one of the orchid AGL6 genes belonging to the class A MADS box genes, PAXXG301780, was expressed throughout all floral organs. Its expression pattern was the only one similar to AtAGL6, but other three AGL6 genes (PAXXG198660, PAXXG228490 and PAXXG348560) did not exhibit this

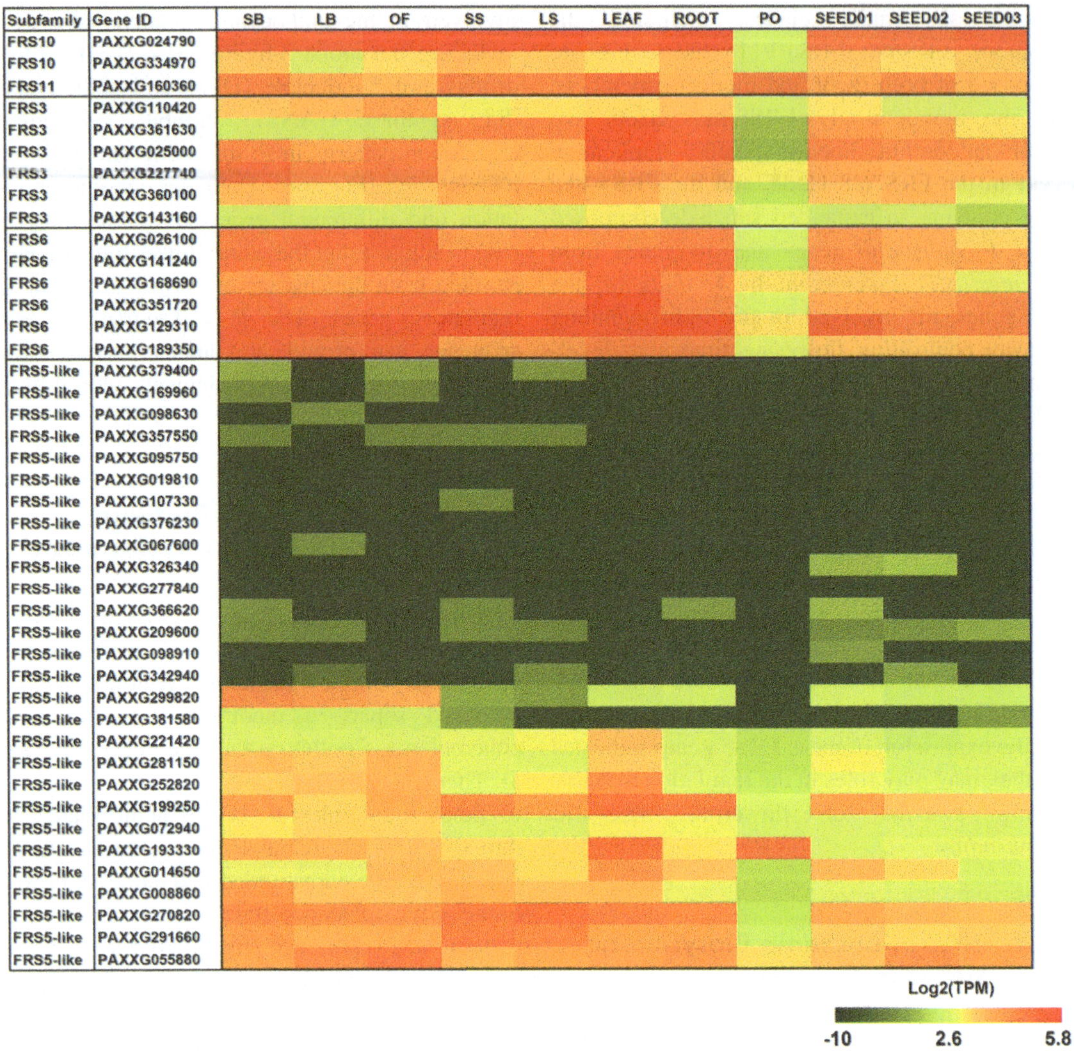

Fig. 8.3 Heatmap for orchid *FAR1/FRS* genes (Adapted from Chao et al. 2018)

conventional expression pattern. Furthermore, PAXXG198660 was specifically expressed in the lip; this may indicate PAXXG198660 plays an essential role in lip formation. The other three A class MADS box genes belonging to AP1/FUL clade in *P. aphrodite* were expressed mainly in the inner whorls of the pollinia and pedicel, indicating that its function may be similar to the function of the FUL gene but not AP1 that are required for maintaining the homeostasis of whorl 1 and 2 (sepal and petal). The previous study supports that AP1 and FUL-like proteins are both in dicot species, whereas monocot species only have FUL-like proteins (Kater et al. 2006). Therefore, orchid AP1/FUL genes might play a role to promote the development of the pollinia and gynoecium but not regulate perianth formation.

In orchid B class MADS box genes, the expression patterns of the five genes (four AP3 and one PI) were all divergent from the typical expression in Arabidopsis (Fig. 8.5b). The first pattern, PAXXG349010 was predominantly expressed in sepals and petals, whorls 1 and 2. The second pattern, represented by PAXXG113000 and PAXXG070630, has high

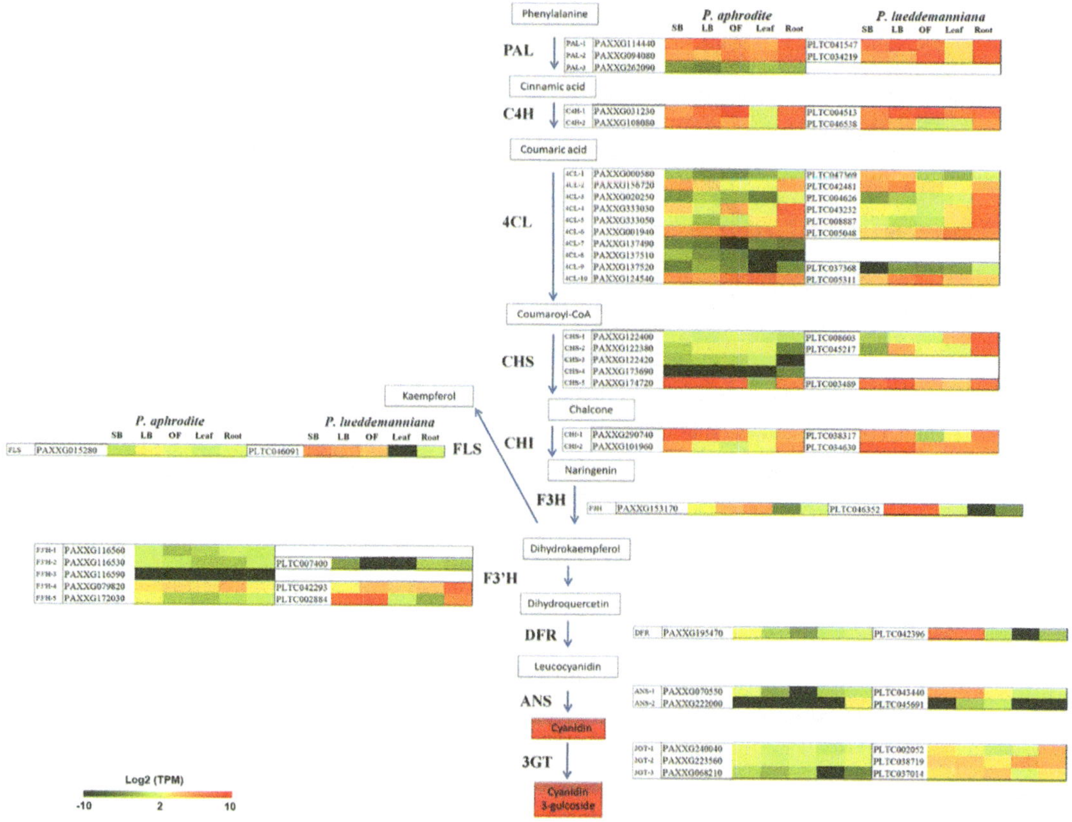

Fig. 8.4 Flavonoid biosynthetic pathway in *P. aphrodite* and *P. lueddemanniana*. Gene expression heatmap of orthologous gene pairs: *P. aphrodite* and *P. lueddemanniana* are shown on the left and right, respectively. LB, large bud; SB, small bud; OF, Open flower. The enzymes involved are indicated next to the blue arrows (Adapted from Chao et al. 2018)

level of expression in the lip and column. The third pattern, represented by PAXXG193450 and PAXXG093380, has high level of expression in all flower organs except low expression the pollinia. Typical B class genes are predominantly expressed in petal and stamen. Three B class genes (PAXXG349010, PAXXG193450 and PAXXG093380) in *P. aphrodite* extending their expression in sepal may indicate their function associated with sepal petalization.

C/D and E class genes in Arabidopsis and *P. aphrodite* may be functional conserved since their expression profiles are similar to each other (Fig. 8.5b). Taken together, our results revised an ABCDE model for orchids to match their unique flower morphogenesis.

8.4 Orchid miRNA Families, Precursors and Target Genes

Some studies have been performed to identify microRNAs and pre-miRNAs from deep sequencing data in orchid species (Aceto et al. 2014; Chao et al. 2014; An et al. 2011), and the expression of some the candidate microRNAs was confirmed by stem-loop real-time RT-PCR. We have identified 181 known miRNAs (which can be classified into 88 miRNA families, Fig. 8.6) and 23 new miRNAs, as well as the complementary strands of the identified miRNAs from four small RNA libraries of *P. aphrodite* (Chao et al. 2014). Many miRNAs are expressed in a tissue-specific manner. For example,

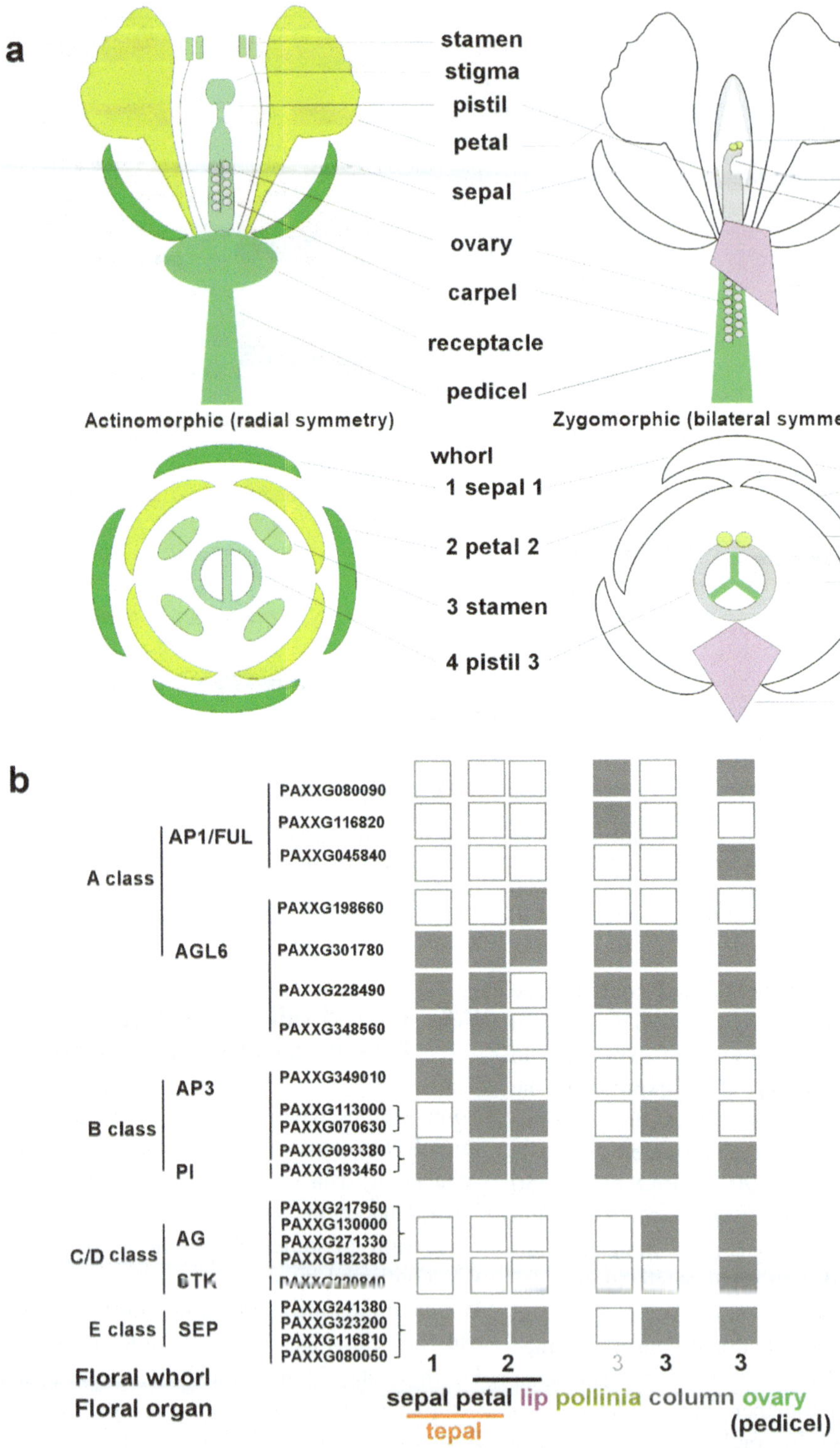

Fig. 8.5 **a** A comparison of flower development in actinomorphic and zygomorphic angiosperms. **b** The MADS box gene expression profiling in *P. aphrodite* (Adapted from Su et al. 2013b)

Table 8.2 A list of MADS-Box genes in *P. aphrodite*

Type	Group	Class	Clade	Genome ID
II	MIKC^C	B	AP3	PAXXG349010
				PAXXG093380
				PAXXG113000
				PAXXG070630
II	MIKC^C	B	PI	PAXXG193450
II	MIKC^C	A	AP1/FUL	PAXXG080090
				PAXXG045840
				PAXXG116820
II	MIKC^C	A	AGL6	PAXXG198660
				PAXXG301780
				PAXXG228490
				PAXXG198670
				PAXXG348560
II	MIKC^C	C/D	AG	PAXXG130000
				PAXXG271330
				PAXXG182380
				PAXXG220840
				PAXXG217950
				PAXXG220800
II	MIKC^C	E	SEP	PAXXG241380
				PAXXG323200
				PAXXG116810
				PAXXG080050
				PAXXG236780
				PAXXG328320
II	MIKC^C		SOC	PAXXG057830
				PAXXG065170
II	MIKC^C		SVP	PAXXG194760
				PAXXG314420
II	MIKC^C		ANR	PAXXG350320
				PAXXG049240
II	MIKC^C		Bs	PAXXG008810
II	MIKC^C		CFO	PAXXG063360
II	MIKC*			PAXXG335970
				PAXXG185550
				PAXXG123370
				PAXXG016180
				PAXXG220830

(continued)

Table 8.2 (continued)

Type	Group	Class	Clade	Genome ID
I	Malpha			PAXXG093020
				PAXXG101820
				PAXXG051820
				PAXXG152520
				PAXXG152330
				PAXXG187720
				PAXXG335930
				PAXXG335935
				PAXXG335905
				PAXXG289610
I	Mgamma			PAXXG289580
				PAXXG113810
				PAXXG201910
				PAXXG026260
I	Mbeta			PAXXG152470
				PAXXG374144
				PAXXG126305
				PAXXG351395

miR2911 is specifically expressed in roots. miR169, miR172, miR395 and miR858 are highly expressed in flowers, miR398 and miR396 are leaf-specific. The miR156 is expressed at higher level in seeds and at very low level in flowers, which is opposite to the expression pattern of miR172.

Plant pre-miRNAs exhibit greater variation in stem-loop size/structure than animal pre-miRNAs, making the prediction of plant pre-miRNAs from sequencing data more challenging. We established a miRNA precursor prediction procedure taking into account specific factors that affect the precursor structural features, including matches, gaps, and bulges on the stem region, the occurrence of multi-loops, and minimum free energy (MFE) (Chao et al. 2014). The miRNA precursor prediction procedure was tested using Arabidopsis pre-miRNA sequences (miRBase version 18) and over 300 negative examples, which resulted in a sensitivity of 85% and a specificity of 100%. Furthermore, we applied the prediction procedure to identify a total of 91 *P. aphrodite* precursors from small RNA sequencing data and transcriptome assembly.

We took both homology-dependent and homology-independent computational approaches to the identification of target genes. The homology-dependent approach relied on finding homologs of known target genes in model organisms. In homology-independent approach, the miRNA to EST alignment with complementary matches was checked, and an EST was considered to be a miRNA target candidate if the alignment contains positions 2–12 of the miRNA with an alignment length over 16 nt and less than or equal to three mismatches. A total of 228 target transcripts were predicted using these two approaches. Moreover, RLM 5′ RACE assay was performed to confirm the cleavage of predicted target transcripts for several miRNAs. After the completion of the *Phalaenopsis* genome sequencing and annotation, we removed the redundancies and revised the targets to 163 non-redundant gene loci corresponding to 41 miRNA families (Table 8.3).

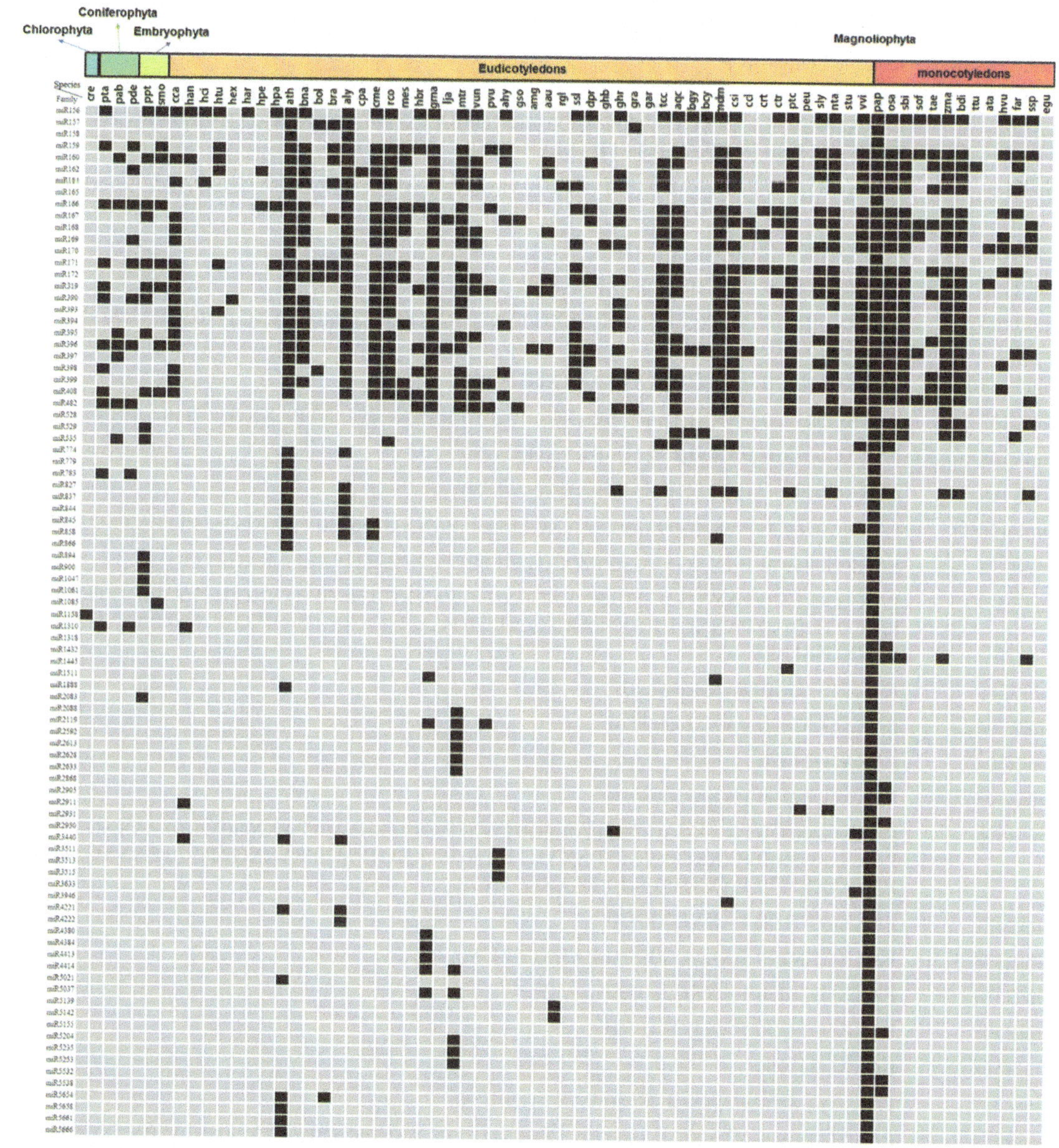

Fig. 8.6 The 88 known miRNA families identified in *P. aphrodite* and their distribution across plant species (Adapted from Chao et al. 2014)

All the mature miRNAs and miRNA precursor data can be found in the Orchidstra 2.0 database (Chao et al. 2017), including the sequence, the predicted hairpin structure of precursor, and location information of the mature miRNA. Each mature miRNA entry is linked with its precursor entry and target protein-coding entries in the Orchidstra 2.0 database.

8.5 Orchidstra 2.0—Genomics and Functional Genomics Database for the Orchid Family

Orchid genomes are usually considerably larger than those of most model organisms, this hampers the progress of genomic resources

Table 8.3 Candidate target genes of the miRNAs in *P. aphrodite*

miR Family	Identified target loci
miR156/157	PAXXG001260 PAXXG002320 PAXXG018830 PAXXG023380 PAXXG040650 PAXXG054010 PAXXG088850 PAXXG130960 PAXXG165750 PAXXG201790 PAXXG219700 PAXXG236750 PAXXG324040 PAXXG328340 PAXXG346560
miR159	PAXXG017110 PAXXG020210 PAXXG034640 PAXXG063440 PAXXG134970 PAXXG149150 PAXXG226520 PAXXG240250 PAXXG243400 PAXXG281070 PAXXG341100 PAXXG363580
miR160	PAXXG146200 PAXXG285910
miR162	PAXXG095200 PAXXG116000 PAXXG321950
miR164	PAXXG007690 PAXXG029620 PAXXG072060 PAXXG123380 PAXXG171030 PAXXG207970 PAXXG250220
miR165/166	PAXXG120220 PAXXG152850 PAXXG209270
miR167	PAXXG023320 PAXXG059450 PAXXG132010 PAXXG291450 PAXXG328840
miR168	PAXXG007510 PAXXG022755 PAXXG049780
miR169	PAXXG049090 PAXXG176040 PAXXG241750
miR170/171	PAXXG076420 PAXXG093750 PAXXG114650 PAXXG166330 PAXXG286860 PAXXG350070
miR172	PAXXG084990 PAXXG130680 PAXXG201440 PAXXG221920 PAXXG242400
miR319	PAXXG016320 PAXXG017110 PAXXG020210 PAXXG034640 PAXXG043760 PAXXG056230 PAXXG110140 PAXXG121830 PAXXG221270 PAXXG243080 PAXXG281070 PAXXG297730 PAXXG341100
miR390	JI766713 (GenBank accession)
miR393	PAXXG053640 PAXXG099820
miR394	PAXXG351570
miR395	PAXXG049770 PAXXG088080
miR396	PAXXG000850 PAXXG013740 PAXXG042670 PAXXG045240 PAXXG052010 PAXXG063600 PAXXG082040 PAXXG082810 PAXXG119150 PAXXG159420 PAXXG224930 PAXXG242510 PAXXG292870 PAXXG314410 PAXXG332670
miR397	PAXXG078090 PAXXG306950
miR398	PAXXG124790 PAXXG207360 PAXXG261790
miR399	PAXXG228230 PAXXG280730
miR408	PAXXG013390 PAXXG341760
miR528	PAXXG248660
miR529	PAXXG018830 PAXXG023380 PAXXG201790
miR535	PAXXG013990 PAXXG033300 PAXXG058480 PAXXG062520 PAXXG085590 PAXXG088700 PAXXG108390 PAXXG108560 PAXXG116660 PAXXG125940 PAXXG143370 PAXXG153800 PAXXG158580 PAXXG185300 PAXXG295280 PAXXG299270 PAXXG341670
miR783	PAXXG178670
miR827	PAXXG291950
miR858	PAXXG050100 PAXXG165560
miR1318	PAXXG144880
miR2628	PAXXG298610
miR2950	PAXXG033300
miR4384	PAXXG150540
miR5021	PAXXG004600 PAXXG019710 PAXXG026110 PAXXG035670 PAXXG045690 PAXXG047840 PAXXG087140 PAXXG137750 PAXXG145410 PAXXG153120 PAXXG166260 PAXXG209280 PAXXG242270 PAXXG278345 PAXXG335680

(continued)

Table 8.3 (continued)

miR Family	Identified target loci
miR5139	PAXXG055220
miR5204	PAXXG141220
miR5658	PAXXG017210 PAXXG140830
PA-miR1-5p	PAXXG070630 PAXXG113000
PA-miR2-3p	PAXXG049240 PAXXG350320
PA-miR3-5p	PAXXG174490
PA-miR7-3p	PAXXG316770
PA-miR14-3p	PAXXG297020
PA-miR16-5p	PAXXG084240

development in orchid species. Even now when the *Phalaenopsis* orchid genomes are available, whole transcriptome sequencing is still the major way to access gene information in other orchid species. We have constructed the Orchidstra 2.0 database (Chao et al. 2017) to serve as a web-based functional genomics resource of the orchid family for the research community.

In order to build the Orchidstra 2.0 database, we sequenced the transcriptomes of various tissues for nine species, and obtained EST data and RNA-Seq data of another nine species from NCBI. We have built several bioinformatics pipelines for the analysis of both coding and non-coding RNA. A total of 510,947 protein-coding genes and 161,826 non-coding transcripts from 18 orchid species, belonging to 12 genera in five subfamilies of Orchidaceae, were stored in the Orchidstra 2.0 database (Fig. 8.7a). The Orchidstra 2.0 database also contains microarray and RNA-Seq gene expression data and provides visualization and exploration tools (Fig. 8.7b).

Identification of orthologous groups is crucial in transferring the acquired functional annotations between orchid species and other well-studied genomes. For this reason, we provide precomputed orthologous groups in 18 orchid species and two model organisms (Arabidopsis and rice) in the Orchidstra 2.0 database, which is useful for gene annotation and comparative genomics. After the completion of *P. aphrodite* genome sequencing and annotation, in order to support the effective utilization of genomic data and connect the genome annotation to tissue-specific gene expression profile, we have created the Genome Resources website to add the *P. aphrodite* genomic assembly and annotation into the Orchidstra 2.0 web database.

The Orchidstra 2.0 database is continually maintained and updated. The database can be searched by sequence similarity, gene ID or by combining keywords using Boolean operators. Gene sequences, ortholog groups, functional annotations and gene expression data are also available for download (Fig. 8.7c). The Orchidstra 2.0 database is available at the following address: http://orchidstra2.abrc.sinica.edu.tw.

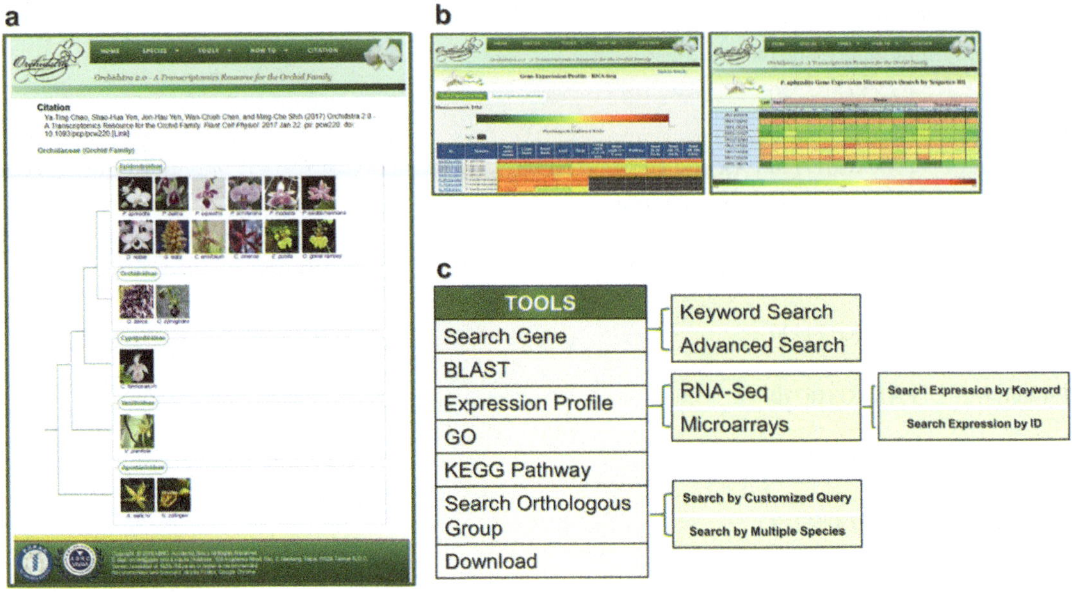

Fig. 8.7 Home page and web tools of Orchidstra 2.0 database. **a** List of orchid species in the database. **b** An example of searching orchid gene expression data. **c** Tools for searching and browsing the database

8.6 Future Prospects

The high-quality orchid reference genome and annotation would provide researchers a great opportunity to look into the genomic features in orchid species. The major challenge of post-genomics research in orchids is the integration of multiomics data, such as metabolomics data, RNA-Seq, epigenomics data, to reveal the modulation elements for horticultural important phenotype or ecological traits. We believe that the Orchidstra 2.0 database provides a good start points for identifying important gene families and genetic variation and will facilitate the development of new breeding platforms for orchid improvement.

References

Aceto S, Sica M, De Paolo S, D'Argenio V, Cantiello P, Salvatore F, Gaudio L (2014) The analysis of the inflorescence miRNome of the orchid *Orchis italica* reveals a DEF-like MADS-box gene as a new miRNA target. PLoS ONE 9(5):e97839. https://doi.org/10.1371/journal.pone.0097839

An FM, Hsiao SR, Chan MT (2011) Sequencing-based approaches reveal low ambient temperature-responsive and tissue-specific microRNAs in *Phalaenopsis* orchid. PLoS ONE 6(5):e18937. https://doi.org/10.1371/journal.pone.0018937

Boetzer M, Henkel CV, Jansen HJ, Butler D, Pirovano W (2011) Scaffolding pre-assembled contigs using SSPACE. Bioinformatics 27(4):578–579. https://doi.org/10.1093/bioinformatics/btq683

Chao YT, Chen WC, Chen CY, Ho HY, Yeh CH, Kuo YT, Su CL, Yen SH, Hsueh HY, Yeh JH, Hsu HL, Tsai YH, Kuo TY, Chang SB, Chen KY, Shih MC (2018) Chromosome-level assembly, genetic and physical mapping of *Phalaenopsis aphrodite* genome provides new insights into species adaptation and resources for orchid breeding. Plant Biotechnol J 16(12):2027–2041. https://doi.org/10.1111/pbi.12936

Chao YT, Su CL, Jean WH, Chen WC, Chang YC, Shih MC (2014) Identification and characterization of the microRNA transcriptome of a moth orchid *Phalaenopsis aphrodite*. Plant Mol Biol 84(4–5):529–548. https://doi.org/10.1007/s11103-013-0150-0

Chao YT, Yen SH, Yeh JH, Chen WC, Shih MC (2017) Orchidstra 2.0-A transcriptomics resource for the orchid family. Plant Cell Physiol 58 (1):e9. https://doi.org/10.1093/pcp/pcw220

Christenson EA (2001) Phalaenopsis: a Monograph. Timber Press, Portland, Oregon

Gnerre S, Maccallum I, Przybylski D, Ribeiro FJ, Burton JN, Walker BJ, Sharpe T, Hall G, Shea TP, Sykes S, Berlin AM, Aird D, Costello M, Daza R, Williams L, Nicol R, Gnirke A, Nusbaum C,

Lander ES, Jaffe DB (2011) High-quality draft assemblies of mammalian genomes from massively parallel sequence data. Proc Natl Acad Sci USA 108 (4):1513–1518. https://doi.org/10.1073/pnas.1017351108

Hsu, CC, Chen, HH, Chen, WH (2018) Phalaenopsis. In: Van Huylenbroeck J (ed) Ornamental Crops. Handbook of Plant Breeding, vol 11. Springer, Cham, pp 567–625

Kater MM, Dreni L, Colombo L (2006) Functional conservation of MADS-box factors controlling floral organ identity in rice and Arabidopsis. J Exp Bot 57 (13):3433–3444. https://doi.org/10.1093/jxb/erl097

Lin R, Wang H (2004) Arabidopsis FHY3/FAR1 gene family and distinct roles of its members in light control of Arabidopsis development. Plant Physiol 136(4):4010–4022. https://doi.org/10.1104/pp.104.052191

Mendoza L, Thieffry D, Alvarez-Buylla ER (1999) Genetic control of flower morphogenesis in *Arabidopsis thaliana*: a logical analysis. Bioinformatics 15(7–8):593–606. https://doi.org/10.1093/bioinformatics/15.7.593

Simao FA, Waterhouse RM, Ioannidis P, Kriventseva EV, Zdobnov EM (2015) BUSCO: assessing genome assembly and annotation completeness with single-copy orthologs. Bioinformatics 31(19):3210–3212. https://doi.org/10.1093/bioinformatics/btv351

Su CL, Chao YT, Yen SH, Chen CY, Chen WC, Chang YC, Shih MC (2013a) Orchidstra: an integrated orchid functional genomics database. Plant Cell Physiol 54(2):e11. https://doi.org/10.1093/pcp/pct004

Su CL, Chen WC, Lee AY, Chen CY, Chang YC, Chao YT, Shih MC (2013b) A modified ABCDE model of flowering in orchids based on gene expression profiling studies of the moth orchid *Phalaenopsis aphrodite*. PLoS ONE 8(11):e80462. https://doi.org/10.1371/journal.pone.0080462

Genes and Noncoding RNAs Involved in Flower Development in *Orchis italica*

9

Serena Aceto

Abstract

Orchis italica is one of the most widespread orchid species in Mediterranean regions. During the last 20 years, great efforts based also on next-generation sequencing approaches have been made to understand the molecular mechanisms underlying orchid flower development, and many genes have been identified in *O. italica*. The identified genes include many MADS-box genes involved in the ABCDE model as well as microRNAs that post-transcriptionally regulate their activity. More recently, genes that are potentially involved in the establishment of bilateral symmetry in the flowers of *O. italica* belonging to the TCP and MYB families of transcription factors have been identified. In addition, microRNAs and long noncoding RNAs have been identified as possibly being involved in the complex process of flower development in *O. italica*.

S. Aceto (✉)
Department of Biology, University of Naples Federico II, Naples, Italy
e-mail: serena.aceto@unina.it

9.1 *Orchis italica*

Orchid flowers are known for their beauty and extreme morphological diversification among different species. However, they share a common organization consisting of three outer tepals (first floral whorl), two inner lateral tepals and an inner median tepal referred to as the labellum or lip (second floral whorl). The male and female reproductive tissues are fused to form the column in the innermost whorl. The pollen grains are located at the apex of the column, whereas the ovary, whose maturation is triggered by pollination, is located at the base of the column (Rudall and Bateman 2002).

The temperate terrestrial orchid subtribe Orchidinae comprises approximately 1,800 species. Among these species, *Orchis italica* Poir. (Fig. 9.1) is one of the most widespread in Mediterranean regions, with a distribution area extending northward to Dalmatia and westward to Portugal. It grows on grasslands up to 1,300 m a.s.l. and blooms between March and June. The inflorescences of *O. italica* are dense, with the color flowers ranging from light pink to purple. The pendent lip is deeply tri-lobed and has a cylindrical spur directed downward (Delforge 2005).

Most of the transcriptomic and genomic studies concerning Orchidaceae are focused on tropical orchid lineages (Cai et al. 2015; Lin et al. 2016; Zhang et al. 2016, 2017; Chao et al. 2017,

F.-C. Chen and S.-W. Chin (eds.), *The Orchid Genome*, Compendium of Plant Genomes,
https://doi.org/10.1007/978-3-030-66826-6_9

Fig. 9.1 Inflorescence of *Orchis italica*

2018; Unruh et al. 2018; Yang et al. 2019). The few studies on the transcriptome and miRNome of *O. italica* (Aceto et al. 2014; De Paolo et al. 2014) have provided a valuable resource for the identification of genes and noncoding RNAs (both small and long) expressed in the flower tissues of this species and for comparison of their expression profiles with those of their orthologs in other orchid species belonging to different subfamilies (De Paolo et al. 2015; Valoroso et al. 2017, 2019a). These genes and noncoding RNAs are mainly involved in flower organ formation and in establishing floral bilateral symmetry.

9.2 The ABCDE Model of Flower Development

The complex genetic pathways that control flower development, from the regulation of floral meristem identity to floral organ identity, have been deeply dissected in model species (e.g., *Arabidopsis thaliana*). Since the first proposal of the ABC model, it appeared to provide an elegant and clear explanation of the interactions among homeotic genes that specify floral organ identity (Schwarz-Sommer et al. 1990; Bowman et al. 1991). The initial model described three functional classes of transcription factors responsible for sepal (class A), petal (class A and B), stamen (class B and C) and carpel identity (class C). Two additional classes of transcription factors have since been included in this model: class D, which participates to ovule development (Pinyopich et al. 2003), and class E, responsible for the determination of all floral organs (Pelaz et al. 2000) (Fig. 9.2a). All but one of the transcription factors included in the ABCDE model belong to the MADS-box family (Heijmans et al. 2012). The only exception is the class A gene *APETALA2*, which belongs to the AP2/ERF family of transcription factors (Bowman et al. 1989).

The ABCDE model is quite well conserved and has been verified in many species (Ambrose et al. 2000; Ferrario et al. 2004; Whipple et al. 2004, 2007). However, modifications of the canonical model have been proposed for non-model species (Irish and Litt 2005). For example, in orchids, the expression of the class B genes is expanded to the outermost floral whorl, and this pattern is related to the presence of petaloid sepals (outer tepals) (Kanno et al. 2007; Cantone et al. 2009; Salemme et al. 2011) (Fig. 9.2b). In addition, the orchid *AP3/DEF* genes (class B) play an important role in the formation of the orchid perianth, and their involvement in the diversification of the lateral and median inner tepals (lip) is explained by the “orchid code”, “homeotic orchid tepal” (HOT) and “P-code” models (Mondragon-Palomino and Theissen 2009, Mondragon-Palomino and Theissen 2011; Pan et al. 2011; Hsu et al. 2015).

9.2.1 The *AP2* Gene

The *AP2* gene belongs to the plant-specific AP2/ERF (APETALA2/Ethylene Responsive Factor) family of transcription factors related to different developmental processes. All the AP2 proteins share a highly conserved DNA-binding domain (Weigel 1995; Okamuro et al. 1997) and

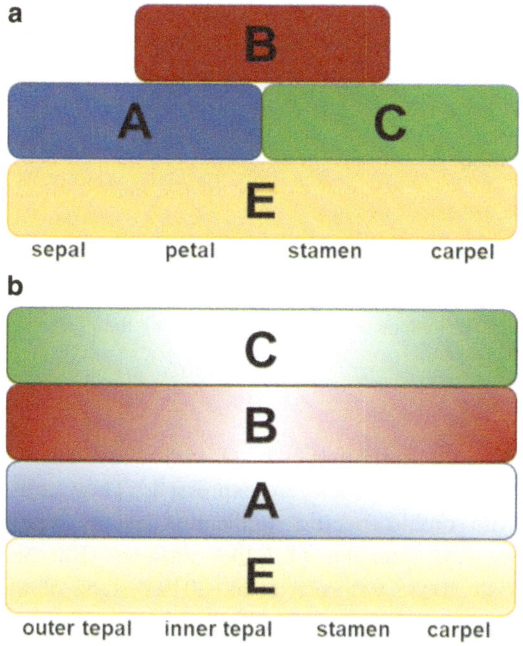

Fig. 9.2 Models of flower organ identity. **a** Canonical ABCE model: class A genes establish the formation of sepals; class A and B genes together specify the formation of petals; class B and C genes drive the formation of stamens; class C genes determine the formation of carpels. Class E genes are necessary for the correct development of all floral organs. **b** Fading border model: the expression domains of the different functional classes of organ identity genes are not clearly defined, and a gradient of expression of the class A–E genes influences the development of petaloid organs (tepals), stamens and carpels. Class D genes, mainly involved in ovary formation, are not reported in this scheme. Modified from Chanderbali et al. (2016)

are encoded by transcripts containing the target site for the microRNA miR172 (Rhoades et al. 2002; Aukerman and Sakai 2003; Chen 2004). In the ABCDE model of flower development, the class A gene *AP2* is involved (either alone or in combination with genes of class B) in the determination of the first and second floral whorls (Kunst et al. 1989; Bowman et al. 1991; Jofuku et al. 1994; Yant et al. 2010). In addition, the *AP2* genes are involved in ovule and seed development (Jofuku et al. 2005; Guillaumot et al. 2008), and some of them are also expressed in vegetative tissues (Licausi et al. 2010; Rashid et al. 2012).

In *O. italica*, the AP2/ERF gene *OitaAP2* includes 10 exons and 9 introns and is subjected to alternative splicing, with the consequent formation of two transcripts produced through differential skipping of exon 9 (Salemme et al. 2013b). Alternative splicing is a post-transcriptional regulation mechanism shared by other *AP2/ERF* genes of *Arabidopsis*, kiwifruit (Varkonyi-Gasic et al. 2012) and peanut (Park and Grabau 2016), and in some cases (e.g., in kiwifruit), the alternative isoform lacks the miR172 target site. In *O. italica*, both *OitaAP2* splicing isoforms preserve the correct reading frame and the target site of miR172. They are expressed in the perianth tissues (outer and inner tepals, lip), ovary and vegetative tissues, whereas they are absent in the fused reproductive tissues (column). Different transcript levels of the two isoforms suggest their possible functional diversification in the development of the perianth organs, in ovary formation before pollination and in vegetative tissues (root and stem) (Salemme et al. 2013b).

miR172 is a negative regulator of *AP2* in *Arabidopsis*, as it recognizes and cleaves a specific target site located in the coding sequence of the *AP2* transcript (Aukerman and Sakai 2003; Chen 2004). Both *OitaAP2* isoforms of *O. italica* are cleaved by miR172 (Fig. 9.3a), and the expression profile of this microRNA is complementary to that of the *OitaAP2* isoforms. For example, high levels of miR172 are present in the column, where the transcripts of *OitaAP2* are absent. However, this pattern is restricted to the floral tissues in the early developmental stage, whereas after anthesis, miR172 is present at very low levels in the perianth and column, suggesting that miR172 exerts its inhibitory function on *OitaAP2* in the early stages of flower development.

The post-transcriptional regulatory mechanism of miR172 related to *AP2*-like genes is shared by other orchids, e.g., *Phalaenopsis* and *Erycina* (An et al. 2011; Lin et al. 2013), and is conserved in other plant species (Zhu and Helliwell 2011; Varkonyi-Gasic et al. 2012). The presence of miR172 and its targets in taxa

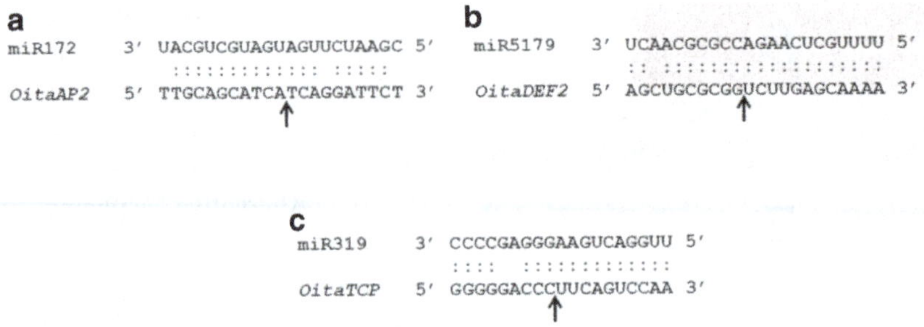

Fig. 9.3 Alignment of miRNAs of *Orchis italica* and their target sites in specific mRNAs. **a** miR172/*OitaAP2* interaction; **b** miR5179/*OitaDEF2* interaction; **c** miR319/*OitaTCP* interaction. The arrows indicate the position of cleavage

ranging from ferns and gymnosperms to flowering plants provides evidence of its ancient origin. During evolution, miR172 has been recruited in different molecular pathways, e.g., the regulation of phase transition (Xing et al. 2014), internode elongation (Patil et al. 2019), and vegetative development (Tang et al. 2018), in addition to the determination of floral organ identity, revealing that the interaction between miR172 and its *AP2/ERF* targets plays an evolutionarily ancient role in plant development.

9.2.2 The MADS-Box Genes

The MADS-box transcription factors are divided into two distinct lineages, type I and II, with different genomic structures, functions and evolutionary histories (Parenicova et al. 2003). Type I MADS-box genes are mainly involved in the development of embryos, seeds and female gametophytes (Masiero et al. 2011), whereas type II genes play roles in different plant molecular pathways, including flower development. In particular, the MADS-box proteins of the ABCDE model of flower development are type II MIKC^{C} transcription factors.

In the inflorescence of *O. italica*, 29 genes encoding MADS-box proteins are expressed (4 type I and 25 type II). Among the type II genes, 19 belong to the MIKC^{C} group and are members of the five functional classes of the ABCDE model of flower development (Valoroso et al. 2019a).

In the floral tissues of *O. italica* four class A MADS-box genes are expressed. They arose from orchid-specific duplication events within the two monocot *FUL*-like lineages (Litt and Irish 2003; Valoroso et al. 2019a). One of the genes lacks the region encoding the FUL-like motif LPPWML at the C-terminus, a characteristic that is also displayed by other orchid *FUL*-like genes of Epidendroideae (Acri-Nunes-Miranda and Mondragon-Palomino 2014) and Apostasioideae. The four *FUL*-like genes of *O. italica* are expressed in all floral tissues, though with different expression profiles. These expression patterns are shared by other orchid *FUL*-like genes (Lin et al. 2016; Dirks-Mulder et al. 2017), suggesting that functional diversification may have occurred after duplication.

In *O. italica,* there are six class B MADS-box genes, two *PI/GLO* and four *AP3/DEF* (Aceto et al. 2007; Cantone et al. 2009; Aceto et al. 2010; Cantone et al. 2011; Salemme et al. 2011; Aceto et al. 2014). The duplication of the orchid *PI/GLO* gene was a lineage-specific event that took place within the Orchidinae subtribe, being *PI/GLO* a single copy gene in the other orchid subfamilies. The two *PI/GLO* genes of *O. italica* are coexpressed in all the floral organs, showing the highest levels in the lip. They are also expressed in the ovary before pollination, whereas after pollination, which triggers ovary maturation in orchids, the expression of both *PI/GLO* genes decreases. This expression pattern is similar to that observed in *Phalaenopsis equestris* (Tsai et al. 2005), showing that in orchids, the *PI/GLO* genes are negative regulators of ovary development. Additionally, the expression pattern of the four *AP3/DEF* genes of *O. italica* is not restricted to the second and third floral whorls but is expanded to the outermost whorl, where it explains the presence of petaloid tepas. In agreement with the "orchid code" model and its revised versions (Mondragon-Palomino and Theissen 2009, 2011; Pan et al. 2011), the *DEF1* and *2* genes of *O. italica* are mostly expressed in the first whorl (outer tepals), whereas *DEF3* and *4* are expressed in the inner tepals (lateral and median) (Aceto et al. 2014). The activity of the *AP3/DEF* genes of *O. italica* is regulated at the post-transcriptional level by the microRNA miR5179. This microRNA cleaves the transcripts of the *DEF2* and *DEF4* genes at a specific target site within the coding sequence (Fig. 9.3b), whereas it does not cleave the *DEF1* and *DEF3* transcripts (Aceto et al. 2014).

Three class C (*AG*) genes and one class D (*STK*) gene are expressed in the floral tissues of *O. italica* (Salemme et al. 2013a; Valoroso et al. 2019a). All of them are highly expressed in the column and ovary with a pattern that is in agreement with the role of these functional classes of transcription factors in the development of male and female reproductive organs. Similar to other orchid members of the ABCDE model of flower development, the transcriptional profile of one *AG* gene of *O. italica* is extended to floral whorls outside the canonical ones, particularly the first and second whorls.

In most orchids, there are four class E genes (*SEP*) originating from orchid-specific duplication events (Acri-Nunes-Miranda and Mondragon-Palomino 2014; Lin et al. 2016; Dirks-Mulder et al. 2017). In *O. italica* and *Habenaria radiata*, another member of the subfamily Orchidoideae as *O. italica*, only two *SEP* genes have been found, both expressed in all the floral whorls (Mitoma and Kanno 2018; Valoroso et al. 2019a).

In orchids, the development of the perianth is strictly controlled by the combined activity of the *AP3/DEF* (class B) and *AGL6* genes. As proposed by the "P-code" model, the specific protein complexes SP and L, formed by different AP3/DEF and AGL6 proteins, alternatively drive the formation of the tepals and lip (Hsu et al. 2015). The three *AGL6* genes of *O. italica* exhibit similar expression levels in the lip, in contrast to the four *AP3/DEF* genes, which display different expression levels (Aceto et al. 2014). Thus, transcriptional and posttranscriptional regulatory mechanisms, including microRNA regulatory function, tune the levels of the different AP3/DEF proteins, resulting in the formation of a specific protein complex (SP or L) within the specific floral whorl.

Comparison of the expression patterns of the MADS-box genes of *O. italica* involved in the ABCDE model of flower development with those of other orchids shows that in Orchidaceae, a gradient of expression of floral homeotic genes overlapping at the boundaries of the specific domains can explain flower organ development better than the canonical ABCDE model. This "fading borders model" (Fig. 9.2b) has also been proposed for other nonmodel species belonging to basal angiosperms, magnoliids and basal eudicots (Chanderbali et al. 2016).

9.3 Floral Symmetry

Most orchid flowers are zygomorphic (bilaterally symmetric) and resupinated; that is, before flowering, the pedicel and ovary are rotated 180° thus transferring the dorsal structures to the ventral position and vice versa (Rudall and Bateman 2002). Starting from the radially symmetric ancestral condition of the flower, zygomorphy has evolved many times during evolution, and the genetic mechanisms underpinning floral symmetry have been dissected in the snapdragon *Antirrhinum majus*, where the interplay of TCP and MYB transcription factors is responsible for the dorso-ventral asymmetry of the flower (Fig. 9.4) (Coen 1996; Luo et al. 1996; Almeida et al. 1997, 1999; Galego and Almeida 2002). In the dorsal part of the flower, the TCP genes *CYCLOIDEA* (*CYC*) and *DICHOTOMA* (*DICH*) activate the expression of the MYB gene *RADIALIS* (*RAD*) (Luo et al.

1996, 1999; Galego and Almeida 2002; Almeida and Galego 2005; Corley et al. 2005). The MYB genes *DIVARICATA* (*DIV*) and *DIV and RAD Interacting Factor* (*DRIF*) are expressed both in the dorsal and ventral domains of the snapdragon flower. In the ventral domains, the DIV/DRIF protein complex binds DNA and regulates the expression of ventralization genes. In dorsal domains, RAD antagonizes DRIF for binding to DIV, preventing the formation of the DIV/DRIF complex and inhibiting ventralization (Raimundo et al. 2013).

In orchids, the development of flower organs that confer zygomorphy to the orchid flower is regulated by the *AP3/DEF* and *AGL6* MADS-box genes (Mondragon-Palomino and Theissen 2009, 2011; Pan et al. 2011; Hsu et al. 2015; Dirks-Mulder et al. 2017). In addition, recent

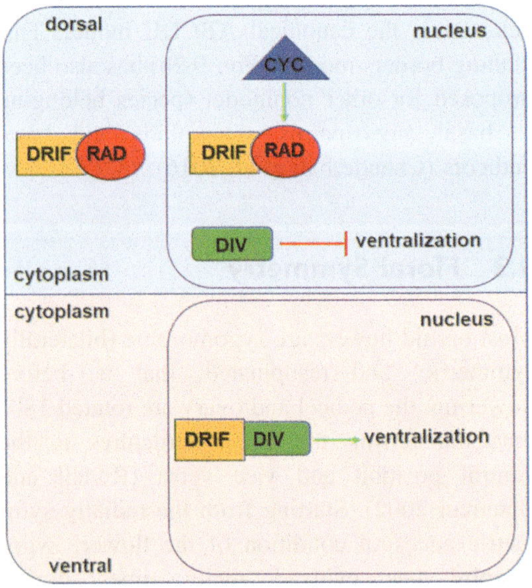

Fig. 9.4 Model of the establishment of bilateral symmetry in *Antirrhinum majus* (modified from Raimundo et al. 2013). In the dorsal domain of the flower, the CYC protein activates the expression of the RAD gene. The DIV and DRIF proteins are present in both the dorsal and ventral domains of the flower. In the dorsal domain, the RAD protein binds to DRIF and inhibits the formation of the DIV/DRIF complex, thus preventing ventralization. In the ventral part of the flower, RAD is not expressed, and DIV can bind to DRIF. The DIV/DRIF protein complex controls the expression of downstream genes involved in ventralization. The green arrows indicate activation, and the red lines indicate inhibition

studies on the TCP and MYB genes have suggested the possible involvement of the *CYC*, *DIV*, *RAD* and *DRIF* genes in the establishment of floral zygomorphy in orchids (Mondragon-Palomino and Trontin 2011; De Paolo et al. 2015; Valoroso et al. 2017; Madrigal et al. 2019).

9.3.1 The TCP Genes

The plant-specific TCP transcription factors control cell proliferation in different developmental processes. Among the TCPs, the *CYC/TB1* genes are involved in the establishment of floral symmetry (Hileman and Cubas 2009; Preston and Hileman 2009; Hileman 2014).

In *O. italica*, 12 TCP genes have been identified (De Paolo et al. 2015). Among these genes, some are ubiquitously expressed and possibly exhibit pleiotropic redundant functions, while others present distinct profiles, suggesting their different roles in specific floral tissues.

One of the 12 TCP genes identified in *O. italica* belongs to the *CYC/TB1* lineage and is weakly expressed in inflorescence tissues. This gene is phylogenetically close to the *OsTB1* gene of *Oryza sativa*, which negatively regulates lateral branching (Takeda et al. 2003).

The microRNA miR319 post-transcriptionally regulates one TCP gene of *O. italica* belonging to the CIN lineage by targeting a cleavage site in its transcript (Fig. 9.3c). This *CIN* gene is mainly expressed in the perianth. The differences in its expression pattern in the inflorescence tissues of *O. italica*, and its regulation by miR319, indicate that it may be involved in the development of the floral tissues of *O. italica* (De Paolo et al. 2015). In *Arabidopsis*, mir319 cleaves a *CIN* transcript and regulates the development of petals and stamens (Nag et al. 2009), suggesting its conserved function in the development of dicot and monocot flower organs.

9.3.2 The MYB Genes

In *O. italica*, there are eight *DIV*, four *RAD* and four *DRIF* genes (Valoroso et al. 2017). These genes are predominantly expressed in the tepals and lip, although some of them are also expressed in other tissues, for example reproductive (column and ovary) and vegetative (leaves).

The orthologs of the *DIV* and *RAD* genes of *A. majus* are expressed in *O. italica* with patterns similar to those observed in snapdragon (Almeida et al. 1997; Galego and Almeida 2002): the transcripts of *DIV* and *DRIF* are present in all the perianth tissues, and those of *RAD* are mainly expressed in the lip (Valoroso et al. 2017). In addition, the in vitro interaction abilities of the DIV, RAD and DRIF proteins of *O. italica* are conserved, with DRIF being able to bind either DIV or RAD, whereas DIV and RAD do not interact (Valoroso et al. 2019b). The expression profile and protein-binding ability of DIV, RAD and DRIF of *O. italica* are in agreement with the interaction module that drives the establishment of bilateral symmetry in *A. majus*: in the lip, the interaction of RAD and DRIF inhibits the formation of the DIV/DRIF complex, preventing ventralization. In other perianth tissues, the DIV and DRIF proteins can interact and activate ventralization. The apparent rotation of the dorsal and ventral expression patterns between *O. italica* and *A. majus* is due to resupination of the orchid flower, with consequent inversion of the dorsal and ventral structures.

9.4 Small and Long Noncoding RNAs

In the inflorescences of *O. italica,* conserved and novel microRNAs have been identified by high-throughput sequencing (Aceto et al. 2014). Their size distribution ranges between 21 and 24 nucleotides, with 24 nt microRNAs showing the highest abundance. Among them, 209 microRNAs of *O. italica* are homologs of known plant microRNAs, and the most highly expressed are miR166, miR168, miR319, miR393, miR396 and miR6300. Almost all of these microRNAs are also highly expressed in other orchids (e.g., *Phalaenopsis*) (An et al. 2011). All eight microRNA families involved in flower development (Luo et al. 2013) are expressed in the inflorescences of *O. italica*, revealing that the regulatory role of microRNAs has been conserved during the evolution of the molecular pathway of flower development.

In addition to miR172, miR5179 and miR319, described above as post-transcriptional regulators of *AP2*, *DEF* and TCP targets, other miRNAs of *O. italica* have putative targets that suggest roles in flower development. For example, miR166 is potentially involved in organ polarity and meristem formation, miR160 in inflorescence and floral organ shaping, and miR396 in flower and leaf development. Another microRNA with a possible role in flower development of *O. italica* is miR169, whose expression is high in tepals and low in the column and ovary. The predicted targets of miR169 are *NF-YA* mRNAs encoding transcription factors that positively regulate the expression of the class C MADS-box gene *AG* (Hong et al. 2003). In *O. italica*, the expression pattern of miR169 is complementary to that of *AG*, demonstrating that the indirect ability of this microRNA to restrict the expression of *AG* to the column and ovary is conserved among plants.

Other microRNAs of *O. italica* have putative targets predicted by in silico analysis and are involved in different biological processes. For example, miR390 of *O. italica* has a putative target site in the homolog of the *TAS3* transcript. In *Arabidopsis,* the interaction between miR390 and *TAS3* mRNA is involved in the regulation of the biogenesis of trans-acting silencing RNAs (ta-siRNAs), which are inhibitors of auxin response factors ARF2-4, involved in the plant response to auxin (Jouannet et al. 2012). Other miRNAs of *O. italica* are involved in the biogenesis or activity of small regulatory RNAs: miR162 has a target site in the *Dicer1* transcript and miR168 in the *AGO1* transcript, which encodes a core component of the RISC complex.

Putative targets have also been identified for novel microRNAs of *O. italica*, such as transcripts encoding transmembrane and leucin-rich repeat protein kinases (Aceto et al. 2014).

The increasing availability of orchid RNA-seq data and the development of ad hoc analysis tools have improved the ability to predict the presence of long noncoding RNAs (lncRNAs) within the assembled transcriptomes of orchid species. Plant lncRNAs are regulators of gene expression, and their function affects different biological functions, from development to responses to stress, through the regulation of chromatin modifications and the mediation of RNA–RNA interactions (Matsui and Seki 2019). Although the genome of *O. italica* is still not available, two software packages, CPC (Kong et al. 2007) and Portrait (Arrial et al. 2009), allowed the identification of 7,779 assembled and unannotated transcripts within the inflorescence transcriptome of *O. italica* matching the criteria for lncRNA attribution. Among these transcripts, some may have a function in male and female reproductive tissues (column), while others may play roles in specific perianth tissues (e.g., lip, outer or inner lateral tepals). In addition, within the inflorescences of *O. italica*, the lncRNA *TAS3* is expressed, which is precursor of several ta-siRNAs and a target of the microRNA miR390, as previously described. miR390 is the phase-initiator microRNA that cleaves *TAS3* at two conserved sites. After cleavage, *TAS3* is converted into double-strand RNA by RDR6, an RNA-dependent RNA polymerase. This double-stranded *TAS3* RNA is then cut into small fragments of 21 nt by DCL4, a Dicer-like enzyme. The produced ta-siRNAs are able to bind the AGO proteins and drive the cleavage of auxin response factor mRNAs (De Paolo et al. 2014).

9.5 Conclusion

Advances in transcriptome/genome analysis and data mining have led to the elucidation of many molecular mechanisms that drive the development of orchid flowers, allowing the proposal of orchid-specific models (e.g., the orchid code model) to explain the peculiar aspects of orchid flowers. The next challenge will be the application of genome editing techniques to orchids, opening the door to functional genetic studies in this nonmodel plant family.

References

Aceto S, Montieri S, Sica M, Gaudio L (2007) Molecular evolution of the *OrcPI* locus in natural populations of Mediterranean orchids. Gene 392:299–305

Aceto S, Cantone C, Chiaiese P, Ruotolo G, Sica M, Gaudio L (2010) Isolation and phylogenetic footprinting analysis of the 5'-regulatory region of the floral homeotic gene *OrcPI* from *Orchis italica* (Orchidaceae). J Hered 101:124–131

Aceto S, Sica M, De Paolo S, D'Argenio V, Cantiello P, Salvatore F, Gaudio L (2014) The analysis of the inflorescence mirnome of the orchid *Orchis italica* reveals a *DEF*-like mads-box gene as a new miRNA target. PLoS ONE 9:e97839

Acri-Nunes-Miranda R, Mondragon-Palomino M (2014) Expression of paralogous *SEP*-, *FUL*-, *AG*- and *STK*-like mads-box genes in wild-type and peloric *Phalaenopsis* flowers. Front Plant Sci 5:76

Almeida J, Galego L (2005) Flower symmetry and shape in *Antirrhinum*. Int J Dev Biol 49:527–537

Almeida J, Rocheta M, Galego L (1997) Genetic control of flower shape in *Antirrhinum majus*. Development 124:1387–1392

Ambrose BA, Lerner DR, Ciceri P, Padilla CM, Yanofsky MF, Schmidt RJ (2000) Molecular and genetic analyses of the *SILKY1* gene reveal conservation in floral organ specification between eudicots and monocots. Mol Cell 5:569–579

An FM, Hsiao SR, Chan MT (2011) Sequencing-based approaches reveal low ambient temperature-responsive and tissue-specific microRNAs in *Phalaenopsis* orchid. PLoS ONE 6:e18937

Arrial RT, Togawa RC, Brigido MM (2009) Screening non-coding RNAs in transcriptomes from neglected species using portrait: case study of the pathogenic fungus *Paracoccidioides brasiliensis*. BMC Bioinform 10:239

Aukerman MJ, Sakai H (2003) Regulation of flowering time and floral organ identity by a microRNA and its *APETALA2*-like target genes. Plant Cell 15:2730–2741

Bowman JL, Smyth DR, Meyerowitz EM (1989) Genes directing flower development in *Arabidopsis*. Plant Cell 1:37–52

Bowman JL, Smyth DR, Meyerowitz EM (1991) Genetic interactions among floral homeotic genes of *Arabidopsis*. Development 112:1–20

Cai J, Liu X, Vanneste K, Proost S, Tsai WC, Liu KW, Chen LJ, He Y, Xu Q, Bian C, Zheng Z, Sun F, Liu W, Hsiao YY, Pan ZJ, Hsu CC, Yang YP, Hsu YC, Chuang YC, Dievart A, Dufayard JF, Xu X, Wang JY, Wang J, Xiao XJ, Zhao XM, Du R, Zhang GQ, Wang M, Su YY, Xie GC, Liu GH, Li LQ, Huang LQ, Luo YB, Chen HH, Van de Peer Y, Liu ZJ (2015) The genome sequence of the orchid *Phalaenopsis equestris*. Nat Genet 47:65–72

Cantone C, Gaudio L, Aceto S (2011) The *PI/GLO*-like locus in orchids: duplication and purifying selection at synonymous sites within Orchidinae (Orchidaceae). Gene 481:48–55

Cantone C, Sica M, Gaudio L, Aceto S (2009) The *OrcPI* locus: genomic organization, expression pattern, and noncoding regions variability in *Orchis italica* (Orchidaceae) and related species. Gene 434:9–15

Chanderbali AS, Berger BA, Howarth DG, Soltis PS, Soltis DE (2016) Evolving ideas on the origin and evolution of flowers: New perspectives in the genomic era. Genetics 202:1255–1265

Chao YT, Yen SH, Yeh JH, Chen WC, Shih MC (2017) Orchidstra 2.0-a transcriptomics resource for the orchid family. Plant Cell Physiol 58:e9

Chao YT, Chen WC, Chen CY, Ho HY, Yeh CH, Kuo YT, Su CL, Yen SH, Hsueh HY, Yeh JH, Hsu HL, Tsai YH, Kuo TY, Chang SB, Chen KY, Shih MC (2018) Chromosome-level assembly, genetic and physical mapping of *Phalaenopsis aphrodite* genome provides new insights into species adaptation and resources for orchid breeding. Plant Biotechnol J 16:2027–2041

Chen XM (2004) A microRNA as a translational repressor of *APETALA2* in *Arabidopsis* flower development. Science 303:2022–2025

Coen ES (1996) Floral symmetry. EMBO J 15:6777–6788

Corley SB, Carpenter R, Copsey L, Coen E (2005) Floral asymmetry involves an interplay between TCP and MYB transcription factors in *Antirrhinum*. Proc Natl Acad Sci USA 102:5068–5073

De Paolo S, Gaudio L, Aceto S (2015) Analysis of the *TCP* genes expressed in the inflorescence of the orchid *Orchis italica*. Sci Rep 5:16265

De Paolo S, Salvemini M, Gaudio L, Aceto S (2014) *De novo* transcriptome assembly from inflorescence of *Orchis italica*: analysis of coding and non-coding transcripts. PLoS ONE 9:e102155

Delforge P (2005) Guide des orchidées d'Europe, d'Afrique du Nord et du Proche- Orient. Delachaux et Niestlé, Neuchâtel-Paris, Switzerland

Dirks-Mulder A, Butot R, van Schaik P, Wijnands JWPM, van den Berg R, Krol L, Doebar S, van Kooperen K, de Boer H, Kramer EM, Smets EF, Vos RA, Vrijdaghs A, Gravendeel B (2017) Exploring the evolutionary origin of floral organs of *Erycina pusilla*, an emerging orchid model system. BMC Evol Biol 17:89

Ferrario S, Immink RG, Angenent GC (2004) Conservation and diversity in flower land. Curr Opin Plant Biol 7:84–91

Galego L, Almeida J (2002) Role of DIVARICATA in the control of dorsoventral asymmetry in *Antirrhinum* flowers. Genes Dev 16:880–891

Guillaumot D, Lelu-Walter MA, Germot A, Meytraud F, Gastinel L, Riou-Khamlichi C (2008) Expression patterns of *LmAP2L1* and *LmAP2L2* encoding two-APETALA2 domain proteins during somatic embryogenesis and germination of hybrid larch (*Larix* x *marschlinsii*). J Plant Physiol 165:1003–1010

Heijmans K, Morel P, Vandenbussche M (2012) MADS-box genes and floral development: the dark side. J Exp Bot 63:5397–5404

Hileman LC (2014) Trends in flower symmetry evolution revealed through phylogenetic and developmental genetic advances. Philos Trans R Soc Lond B Biol Sci 369:20130348

Hileman LC, Cubas P (2009) An expanded evolutionary role for flower symmetry genes. J Biol 8:90

Hong RL, Hamaguchi L, Busch MA, Weigel D (2003) Regulatory elements of the floral homeotic gene *AGAMOUS* identified by phylogenetic footprinting and shadowing. Plant Cell 15:1296–1309

Hsu HF, Hsu WH, Lee YI, Mao WT, Yang JY, Li JY, Yang CH (2015) Model for perianth formation in orchids. Nat Plants 1:15046

Irish VF, Litt A (2005) Flower development and evolution: Gene duplication, diversification and redeployment. Curr Opin Genet Dev 15:454–460

Jofuku KD, Denboer BGW, Vanmontagu M, Okamuro JK (1994) Control of *Arabidopsis* flower and seed development by the homeotic gene *APETALA2*. Plant Cell 6:1211–1225

Jofuku KD, Omidyar PK, Gee Z, Okamuro JK (2005) Control of seed mass and seed yield by the floral homeotic gene *APETALA2*. Proc Natl Acad Sci USA 102:3117–3122

Jouannet V, Moreno AB, Elmayan T, Vaucheret H, Crespi MD, Maizel A (2012) Cytoplasmic *Arabidopsis* AGO7 accumulates in membrane-associated sirna bodies and is required for ta-siRNA biogenesis. EMBO J 31:1704–1713

Kanno A, Nakada M, Akita Y, Hirai M (2007) Class B gene expression and the modified ABC model in nongrass monocots. Sci World J 7:268–279

Kong L, Zhang Y, Ye ZQ, Liu XQ, Zhao SQ, Wei L, Gao G (2007) CPC: Assess the protein-coding potential of transcripts using sequence features and support vector machine. Nucleic Acids Res 35:W345–349

Kunst L, Klenz JE, Martinezzapater J, Haughn GW (1989) *AP2* gene determines the identity of perianth organs in flowers of *Arabidopsis thaliana*. Plant Cell 1:1195–1208

Licausi F, Giorgi FM, Zenoni S, Osti F, Pezzotti M, Perata P (2010) Genomic and transcriptomic analysis of the AP2/ERF superfamily in *Vitis vinifera*. BMC Genomics 11:719

Lin CS, Hsu CT, Liao DC, Chang WJ, Chou ML, Huang YT, Chen JJ, Ko SS, Chan MT, Shih MC (2016) Transcriptome-wide analysis of the MADS-box gene family in the orchid *Erycina pusilla*. Plant Biotechnol J 14:284–298

Lin CS, Chen JJW, Huang YT, Hsu CT, Lu HC, Chou ML, Chen LC, Ou CI, Liao DC, Yeh YY, Chang SB, Shen SC, Wu FH, Shih MC, Chan MT (2013) Catalog of *Erycina pusilla* miRNAs and categorization of reproductive phase-related miRNAs and their target gene families. Plant Mol Biol 82:193–204

Litt A, Irish VF (2003) Duplication and diversification in the *APETALA1/FRUITFULL* floral homeotic gene lineage: implications for the evolution of floral development. Genetics 165:821–833

Luo D, Carpenter R, Vincent C, Copsey L, Coen E (1996) Origin of floral asymmetry in *Antirrhinum*. Nature 383:794–799

Luo D, Carpenter R, Copsey L, Vincent C, Clark J, Coen E (1999) Control of organ asymmetry in flowers of *Antirrhinum*. Cell 99:367–376

Luo Y, Guo Z, Li L (2013) Evolutionary conservation of microRNA regulatory programs in plant flower development. Dev Biol 380:133–144

Madrigal Y, Alzate JF, Gonzalez F, Pabon-Mora N (2019) Evolution of *RADIALIS* and *DIVARICATA* gene lineages in flowering plants with an expanded sampling in non-core eudicots. Am J Bot 106:334-351

Masiero S, Colombo L, Grini PE, Schnittger A, Kater MM (2011) The emerging importance of type I MADS box transcription factors for plant reproduction. Plant Cell 23:865–872

Matsui A, Seki M (2019) The involvement of long noncoding RNAs in response to plant stress. Methods Mol Biol 1933:151–171

Mitoma M, Kanno A (2018) The greenish flower phenotype of *Habenaria radiata* (Orchidaceae) is caused by a mutation in the *SEPALLATA*-like MADS-box gene *HrSEP-1*. Front Plant Sci 9:831

Mondragon-Palomino M, Theissen G (2009) Why are orchid flowers so diverse? Reduction of evolutionary constraints by paralogues of class B floral homeotic genes. Ann Bot 104:583–594

Mondragon-Palomino M, Trontin C (2011) High time for a roll call: gene duplication and phylogenetic relationships of TCP-like genes in monocots. Ann Bot 107:1533–1544

Mondragon-Palomino M, Theissen G (2011) Conserved differential expression of paralogous *DEFICIENS*- and *GLOBOSA*-like mads-box genes in the flowers of Orchidaceae: refining the 'orchid code'. Plant J 66:1008–1019

Nag A, King S, Jack T (2009) Mir319a targeting of TCP4 is critical for petal growth and development in *Arabidopsis*. Proc Natl Acad Sci USA 106:22534–22539

Okamuro JK, Caster B, Villarroel R, VanMontagu M, Jofuku KD (1997) The AP2 domain of APETALA2 defines a large new family of DNA binding proteins in *Arabidopsis*. Proc Natl Acad Sci USA 94:7076–7081

Pan ZJ, Cheng CC, Tsai WC, Chung MC, Chen WH, Hu JM, Chen HH (2011) The duplicated B-class MADS-box genes display dualistic characters in orchid floral organ identity and growth. Plant Cell Physiol 52:1515–1531

Parenicova L, de Folter S, Kieffer M, Horner DS, Favalli C, Busscher J, Cook HE, Ingram RM, Kater MM, Davies B, Angenent GC, Colombo L (2003) Molecular and phylogenetic analyses of the complete MADS-box transcription factor family in *Arabidopsis*: new openings to the MADS world. Plant Cell 15:1538–1551

Park SY, Grabau E (2016) Differential isoform expression and protein localization from alternatively spliced *APETALA2* in peanut under drought stress. J Plant Physiol 206:98–102

Patil V, McDermott HI, McAllister T, Cummins M, Silva JC, Mollison E, Meikle R, Morris J, Hedley PE, Waugh R, Dockter C, Hansson M, McKim SM (2019) APETALA2 control of barley internode elongation. Development 146:dev170373

Pelaz S, Ditta GS, Baumann E, Wisman E, Yanofsky MF (2000) B and C floral organ identity functions require *SEPALLATA* MADS-box genes. Nature 405:200–203

Pinyopich A, Ditta GS, Savidge B, Liljegren SJ, Baumann E, Wisman E, Yanofsky MF (2003) Assessing the redundancy of MADS-box genes during carpel and ovule development. Nature 424:85–88

Preston JC, Hileman LC (2009) Developmental genetics of floral symmetry evolution. Trends Plant Sci 14:147–154

Raimundo J, Sobral R, Bailey P, Azevedo H, Galego L, Almeida J, Coen E, Costa MMR (2013) A subcellular tug of war involving three MYB-like proteins underlies a molecular antagonism in *Antirrhinum* flower asymmetry. Plant J 75:527–538

Rashid M, He GY, Yang GX, Hussain J, Yan X (2012) AP2/ERF transcription factor in rice: genome-wide canvas and syntenic relationships between monocots and eudicots. Evol Bioinform 8:321–355

Rhoades MW, Reinhart BJ, Lim LP, Burge CB, Bartel B, Bartel DP (2002) Prediction of plant microRNA targets. Cell 110:513–520

Rudall PJ, Bateman RM (2002) Roles of synorganisation, zygomorphy and heterotopy in floral evolution: The gynostemium and labellum of orchids and other lilioid monocots. Biol Rev 77:403–441

Salemme M, Sica M, Gaudio L, Aceto S (2011) Expression pattern of two paralogs of the *PI/GLO*-like locus during *Orchis italica* (Orchidaceae, Orchidinae) flower development. Dev Genes Evol 221:241–246

Salemme M, Sica M, Gaudio L, Aceto S (2013a) The *OitaAG* and *OitaSTK* genes of the orchid *Orchis italica*: a comparative analysis with other C- and D-class MADS-box genes. Mol Biol Rep 40:3523–3535

Salemme M, Sica M, Iazzetti G, Gaudio L, Aceto S (2013b) The *AP2*-like gene *OitaAP2* is alternatively spliced and differentially expressed in inflorescence and vegetative tissues of the orchid *Orchis italica*. PLoS ONE 8:e77454

Schwarz-Sommer Z, Huijser P, Nacken W, Saedler H, Sommer H (1990) Genetic control of flower development by homeotic genes in *Antirrhinum majus*. Science 250:931–936

Takeda T, Suwa Y, Suzuki M, Kitano H, Ueguchi-Tanaka M, Ashikari M, Matsuoka M, Ueguchi C (2003) The *OsTB1* gene negatively regulates lateral branching in rice. Plant J 33:513–520

Tang M, Bai X, Niu LJ, Chai X, Chen MS, Xu ZF (2018) Mir172 regulates both vegetative and reproductive development in the perennial woody plant *Jatropha curcas*. Plant Cell Physiol 59:2549–2563

Tsai WC, Lee PF, Chen HI, Hsiao YY, Wei WJ, Pan ZJ, Chuang MH, Kuoh CS, Chen WH, Chen HH (2005) *PeMADS6*, a *GLOBOSA/PISTILLATA*-like gene in *Phalaenopsis equestris* involved in petaloid formation, and correlated with flower longevity and ovary development. Plant Cell Physiol 46:1125–1139

Unruh SA, McKain MR, Lee YI, Yukawa T, McCormick MK, Shefferson RP, Smithson A, Leebens-Mack JH, Pires JC (2018) Phylotranscriptomic analysis and genome evolution of the Cypripedioideae (Orchidaceae). Am J Bot 105:631–640

Valoroso MC, De Paolo S, Iazzetti G, Aceto S (2017) Transcriptome-wide identification and expression analysis of *DIVARICATA*- and *RADIALIS*-like genes of the Mediterranean orchid *Orchis italica*. Genome Biol Evol 9:evx101

Valoroso MC, Censullo MC, Aceto S (2019a) The MADS-box genes expressed in the inflorescence of *Orchis italica* (Orchidaceae). PLoS ONE 14:e0213185

Valoroso MC, Sobral R, Saccone G, Salvemini M, Costa MMR, Aceto S (2019b) Evolutionary conservation of the orchid MYB transcription factors DIV, RAD, and DRIF. Front Plant Sci 10:1359

Varkonyi-Gasic E, Lough RH, Moss SMA, Wu RM, Hellens RP (2012) Kiwifruit floral gene *APETALA2* is alternatively spliced and accumulates in aberrant indeterminate flowers in the absence of miR172. Plant Mol Biol 78:417–429

Weigel D (1995) The APETALA2 domain is related to a novel type of DNA-binding domain. Plant Cell 7:388–389

Whipple CJ, Zanis MJ, Kellogg EA, Schmidt RJ (2007) Conservation of B class gene expression in the second whorl of a basal grass and outgroups links the origin of lodicules and petals. Proc Natl Acad Sci U S A 104:1081–1086

Whipple CJ, Ciceri P, Padilla CM, Ambrose BA, Bandong SL, Schmidt RJ (2004) Conservation of B-class floral homeotic gene function between maize and *Arabidopsis*. Development 131:6083–6091

Xing L, Zhang D, Li Y, Zhao C, Zhang S, Shen Y, An N, Han M (2014) Genome-wide identification of vegetative phase transition-associated micrornas and target predictions using degradome sequencing in *Malus hupehensis*. BMC Genomics 15:1125

Yang FX, Zhu GF, Wei YL, Gao J, Liang G, Peng LY, Lu CQ, Jin JP (2019) Low-temperature-induced changes in the transcriptome reveal a major role of *CgSVP* genes in regulating flowering of *Cymbidium goeringii*. BMC Genomics 20:53

Yant L, Mathieu J, Dinh TT, Ott F, Lanz C, Wollmann H, Chen XM, Schmid M (2010) Orchestration of the floral transition and floral development in *Arabidopsis* by the bifunctional transcription factor APETALA2. Plant Cell 22:2156–2170

Zhang GQ, Liu KW, Li Z, Lohaus R, Hsiao YY, Niu SC, Wang JY, Lin YC, Xu Q, Chen LJ, Yoshida K, Fujiwara S, Wang ZW, Zhang YQ, Mitsuda N, Wang M, Liu GH, Pecoraro L, Huang HX, Xiao XJ, Lin M, Wu XY, Wu WL, Chen YY, Chang SB, Sakamoto S, Ohme-Takagi M, Yagi M, Zeng SJ, Shen CY, Yeh CM, Luo YB, Tsai WC, Van de Peer Y, Liu ZJ (2017) The *Apostasia* genome and the evolution of orchids. Nature 549:379–383

Zhang GQ, Xu Q, Bian C, Tsai WC, Yeh CM, Liu KW, Yoshida K, Zhang LS, Chang SB, Chen F, Shi Y, Su YY, Zhang YQ, Chen LJ, Yin Y, Lin M, Huang H, Deng H, Wang ZW, Zhu SL, Zhao X, Deng C, Niu SC, Huang J, Wang M, Liu GH, Yang HJ, Xiao XJ, Hsiao YY, Wu WL, Chen YY, Mitsuda N, Ohme-Takagi M, Luo YB, Van de Peer Y, Liu ZJ (2016) The *Dendrobium catenatum* Lindl. genome sequence provides insights into polysaccharide synthase, floral development and adaptive evolution. Sci Rep 6:19029

Zhu QH, Helliwell CA (2011) Regulation of flowering time and floral patterning by miR172. J Exp Bot 62:487–495

Phylogeny, Polymorphism, and SSR Markers of *Phalaenopsis* 10

Xiao-Lei Jin, Chi-Chu Tsai, Ya-Zhu Ko, and Yu-Chung Chiang

Abstract

Molecular identification is an essential technology that establishes microsatellite loci in order to improve the commercial breeding of the moth orchid (*Phalaenopsis* species). In this chapter, phylogenetic reconstruction revealed the relationship of *Phalaenopsis* species. Microsatellite primer sets, including genomic-SSR and EST-SSR, were generated from *Phalaenopsis aphrodite* subsp. *formosana* to further investigate the transferability across *Phalaenopsis* species. Ten and twenty-eight polymorphic EST-SSRs and genomic-SSRs markers were collected by magnetic bead enrichment method and next-generation sequencing technology, which revealed 21 species of the genus *Phalaenopsis* with high transferability, particularly in the subgenus *Phalaenopsis*. The result indicated that these microsatellite markers are distinct to subgenus *Phalaenopsis*. Therefore, the genetic relationships among species of subgenus *Phalaenopsis* could be differentiated and constructed on the assignment test. These molecular markers can be used to assess the paternity of *Phalaenopsis* and investigate the hybridization among *Phalaenopsis* species.

10.1 Background of *Phalaenopsis*

The genus *Phalaenopsis* is from the family of Orchidaceae, the subfamily of *Epidendroideae*, tribe *Vandeae*, and subtribe *Aeridinae* (Dressler 1993), which is commonly referred to as moth orchid. There are about 66 species (Christenson 2001). *Phalaenopsis* could be divided into five subgenera according to the number of pollinium (Christenson 2001) and molecular evidence (Tsai et al. 2009). There are three pollinia clades of subgenera *Proboscidioides*, *Aphyllae*, and *Parishianae* and two pollinia clades of subgenera *Polychilos* and *Phalaenopsis*. *Polychilos* and *Phalaenopsis* are further subdivided individually into four sections: *Polychilos*, *Fuscatae*, *Amboinenses*, and *Zebrinae* and *Phalaenopsis*, *Deliciosae*, *Esmeralda*, and *Stauroglottis*, respectively (Dressler 1993; Christenson 2001). *Phalaenopsis* species are widely distributed in the Himalayas of Northern India, South India, Sri Lanka, Southeast China, Taiwan, Indonesia, Thailand, Myanmar, Malaysia, Philippines, Papua New Guinea, and northeastern Australia (Chen and Chen 2011; Christenson 2001). Taiwan is a subtropical island located at the southeastern Asian continent; the climate conditions are certainly conducive for orchids to grow there. Lately, Taiwan has become one of the essential orchid

X.-L. Jin · C.-C. Tsai · Y.-Z. Ko · Y.-C. Chiang (✉)
National Sun Yat-Sen University, Kaohsiung, Taiwan
e-mail: yuchung@mail.nsysu.edu.tw

F.-C. Chen and S.-W. Chin (eds.), *The Orchid Genome*, Compendium of Plant Genomes,
https://doi.org/10.1007/978-3-030-66826-6_10

exporters due to the high quality of breeding and micropropagation technology attached with market demands of the orchid genus *Phalaenopsis* Blume (Orchidaceae) (Chen and Chen 2007, 2011; Tang and Chen 2007).

Phalaenopsis is the most attractive epiphytic monopodial orchid because it has distinctive and varied flowers with a rare structure. Nevertheless, the identification of *Phalaenopsis* varieties and the cultivation of potted plants are challenging and time-consuming because of the complicated phenotype and long stage of the juvenile. Additionally, the traditional horticultural breeding techniques for new cultivars of *Phalaenopsis* due to morphology, physiological development, and environmental factors, as well as their complex interactions, make the breeding results unpredictable and uncertain. It is necessary and useful to develop some highly consistent, fast, and inexpensive techniques for distinguishing and categorizing the identity of *Phalaenopsis* species and cultivars in order to improve the breeding efficiency. Molecular markers could be a delicate and correct tool of choice for identifying species and cultivars. Moreover, the development of molecular markers could be applied to paternity testing and phylogenetic reconstruction and to solve long-standing problems in *Phalaenopsis* breeding. Microsatellite markers are useful tools for plant genetics and crop breeding, including fruit tree and orchid with the high levels of polymorphism, significant heritability, and reproducibility (Powell et al. 1996; Chiang et al. 2012; Chiou et al. 2012; Tsai et al. 2013; Lai et al. 2015, Tsai et al. 2014, 2015).

10.2 Phylogenetics of the *Phalaenopsis*

All *Phalaenopsis* species have 38 (2n = 38) chromosomes excluding the natural tetraploid species *P. buyssoniana* Rchb.f. (Christenson 2001; Tanaka and Kamemoto 1984). Lately, the plastid genome of *P. aphrodite* has been completely sequenced (Chang et al. 2006), and the molecular phylogenies of *Phalaenopsis* species also have been constructed based on the internal transcribed spacer (ITS) of the ribosomal (rDNA) and plastid DNA (Tsai et al. 2006, 2010a, b, 2012). Moreover, molecular data were used to resolve the inheritance of the natural hybrid, *P. x intermedia*, which showed that *P. aphrodite* was the maternal parent and *P. equestris* the paternal parent (Tsai et al. 2006). Besides, complete genome sequencing has been directed in *P. equestris* (Cai et al. 2015).

10.3 Molecular Phylogeny of *Phalaenopsis* on the Basis of Plastid and Nuclear DNA

Sweet (1980) separated the *Phalaenopsis* genus into eight parts (Table 10.1), but Shim (1983) did not agree with the concept of Sweet (1980) and organized the sections *Proboscidioides, Aphyllae, Parishianae, Polychilos, Zebrinae, Fuscatae*, and *Amboinenses* as the genus *Polychilos*. However, Christenson (2001) agreed with Sweet's arrangement and proposed five subgenera. Therefore, he arranged the traditional genera *Kingidium* and *Doritis* as synonyms of *Phalaenopsis*. He divided *Kingidium* into different parts of the *Phalaenopsis*, placing some species (i.e., *P. braceana, P. minus*, and *P. taenialis*) into the subgenus *Aphyllae* and some (i.e., *P. chibae* and P. *deliciosa*) into the section Deliciosae of subgenus *Phalaenopsis* (Table 10.1). This generic level of arrangement was sustained by some outlines of molecular evidence (Padolina et al. 2005; Yukawa et al. 2005; Tsai et al. 2006).

Moreover, the two genera *Nothodoritis* and *Lesliea* of the subtribe *Aeridinae* were nested in *Phalaenposis* contingent on molecular data (Topik et al. 2005; Yukawa et al. 2005). Yukawa et al. (2005) divided *Phalaenopsis* into two genera, *Phalaenopsis* and *Doritis,* following the narrow genus concept. The subgenus *Aphyllae* was not monophyletic based on some molecular evidences (Padolina et al. 2005; Yukawa et al. 2005; Tsai et al. 2006) according to the systematics of *Phalaenopsis* by Christenson (2001). It establishes a clade with *P. lowii* of subgenus *Proboscidioides*, and the sections *Esmeralda* and

Table 10.1 The relationship of the systematics of the genus *Phalaenopsis* among Sweet (1980), Shim (1983), and Christenson (2001) (from Tsai et al. 2010a)

Sweet (1980)	Shim (1983)	Christenson (2001)
Genus *Phalaenopsis*	Genus *Polychilos*	Genus *Phalaenopsis*
Section Proboscidioides	**Section *Proboscidioides***	**Subgenus *Proboscidioides***
Phalaenopsis lowii	*Polychilos lowii*	*Phalaenopsis lowii*
Section *Aphyllae*	**Section *Aphyllae***	**Subgenus *Aphyllae***
		Phalaenopsis braceana (syn. *Kingidium braceana*)[b]
		Phalaenopsis minus[a]
Phalaenopsis wilsonii	*Polychilos wilsonii*	*Phalaenopsis wilsonii*
		Phalaenopsis honghenensis[a]
		Phalaenopsis taenialis (syn. *Kingidium taenialis*)[b]
Phalaenopsis stobartiana (syn. *Phalaenopsis hainanensis*)	*Polychilos stobartiana* (syn. *Polychilos hainanensis*)	*Phalaenopsis stobartiana* *Phalaenopsis hainanensis*
Section *Parishianae*	**Section *Parishianae***	**Subgenus *Parishianae***
Phalaenopsis mysorensis	*Polychilos mysorensis*	Move to the section *Deliciosae*
Phalaenopsis appendiculata	*Polychilos appendiculata*	*Phalaenopsis appendiculata*
Phalaenopsis gibbosa	*Polychilos gibbosa*	*Phalaenopsis gibbosa*
Phalaenopsis parishii	*Polychilos parishii*	*Phalaenopsis parishii*
Phalaenopsis lobii	*Polychilos lobii*	*Phalaenopsis lobii*
Section *Polychilos*	**Section *Polychilos***	**Subgenus *Polychilos***
		Section *Polychilos*
Phalaenopsis mannii	*Polychilos mannii*	*Phalaenopsis mannii*
Phalaenopsis cornu-cervi *Phalaenopsis lamelligera*	*Polychilos cornu-cervi* *Polychilos lamelligera*	*Phalaenopsis cornu-cervi*
		Phalaenopsis borneensis[a]
Phalaenopsis pantherina	*Polychilos pantherina*	*Phalaenopsis pantherina*
Section *Fuscatae*	**Section *Fuscatae***	**Section *Fuscatae***
Phalaenopsis cochlearis	*Polychilos cochlearis*	*Phalaenopsis cochlearis*
Phalaenopsis viridis	*Polychilos viridis*	*Phalaenopsis viridis*
Phalaenopsis fuscata	*Polychilos fuscata*	*Phalaenopsis fuscata*
Phalaenopsis kunstleri	*Phalaenopsis kunstleri*	*Phalaenopsis kunstleri*
Section *Amboinenses*	**Section *Amboinenses***	**Section *Amboinenses***
Phalaenopsis micholitzii	*Polychilos micholitzii*	*Phalaenopsis micholitzii*
Phalaenopsis gigantean	*Polychilos gigantea*	*Phalaenopsis gigantea*
Phalaenopsis javanica	*Polychilos javanica*	*Phalaenopsis javanica*
Phalaenopsis amboinensis	*Polychilos amboinensis*	*Phalaenopsis amboinensis*
Phalaenopsis robinsonii	*Polychilos robinsonii*	*Phalaenopsis robinsonii*
		Phalaenopsis venosa[a]
		Phalaenopsis doweryensis[a]

(continued)

Table 10.1 (continued)

Sweet (1980)	Shim (1983)	Christenson (2001)
Section *Zebrinae*	**Section *Zebrinae***	
Subsection *Lueddemanninae*	**Subsection *Lueddemanninae***	
Phalaenopsis pulchra	*Polychilos pulchra*	*Phalaenopsis pulchra*
Phalaenopsis reichenbachiana	*Polychilos reichenbachiana*	*Phalaenopsis reichenbachiana*
Phalaenopsis fasciata	*Polychilos fasciata*	*Phalaenopsis fasciata*
Phalaenopsis fimbriata	*Polychilos fimbriata*	*Phalaenopsis fimbriata*
Phalaenopsis hieroglyphica	*Polychilos hieroglyphica*	*Phalaenopsis hieroglyphica*
Phalaenopsis lueddemanniana	*Polychilos lueddemanniana*	*Phalaenopsis lueddemanniana*
Phalaenopsis violacea	*Polychilos violacea*	*Phalaenopsis violacea*
Phalaenopsis × *gersenii*	*Polychilos* × *gersenii*	*Phalaenopsis* × *gersenii*
(syn. *Phalaenopsis* × *singuliflora*)	(syn. *Polychilos* × *singuliflora*)	*Phalaenopsis* × *singuliflora*
		Phalaenopsis bastianii [a]
		Phalaenopsis floresensis[a]
		Phalaenopsis bellina[b]
Subsection *Zebrinae*	**Subsection *Zebrinae***	**Section *Zebrinae***
Phalaenopsis speciose	*Polychilos speciosa*	*Phalaenopsis speciosa*
Phalaenopsis tetraspis	*Polychilos tetraspis*	*Phalaenopsis tetraspis*
Phalaenopsis corningiana	*Polychilos corningiana*	*Phalaenopsis corningiana*
Phalaenopsis sumatrana	*Polychilos sumatrana*	*Phalaenopsis sumatrana*
		Phalaenopsis inscriptiosinensis[a]
Subsection *Hirsutae*	**Subsection *Hirsutae***	
Phalaenopsis pallens	*Polychilos pallens*	*Phalaenopsis pallens*
Phalaenopsis mariae	*Polychilos mariae*	*Phalaenopsis mariae*
Subsection *Glabrae*	**Subsection *Glabrae***	
Phalaenopsis modesta	*Polychilos modesta*	*Phalaenopsis modesta*
Phalaenopsis maculate	*Polychilos maculata*	*Phalaenopsis maculata*
	Genus *Phalaenopsis*	**Subgenus *Phalaenopsis***
Section *Phalaenopsis*	**Section *Phalaenopsis***	**Section *Phalaenopsis***
		Phalaenopsis philippinensis[a]
Phalaenopsis amabilis	*Phalaenopsis amabilis*	*Phalaenopsis amabilis*
Phalaenopsis aphrodite	*Phalaenopsis aphrodite*	*Phalaenopsis aphrodite*
Phalaenopsis sanderiana	*Phalaenopsis sanderiana*	*Phalaenopsis sanderiana*
Phalaenopsis schilleriana	*Phalaenopsis schilleriana*	*Phalaenopsis schilleriana*
Phalaenopsis stuartiana	*Phalaenopsis stuartiana*	*Phalaenopsis stuartiana*
Phalaenopsis × *amphitrite*	*Phalaenopsis* × *amphitrite*	*Phalaenopsis* × *amphitrite*
Phalaenopsis × *intermedia*	*Phalaenopsis* × *intermedia*	*Phalaenopsis* × *intermedia*
Phalaenopsis × *leucorrhoda*	*Phalaenopsis* × *leucorrhoda*	*Phalaenopsis* × *leucorrhoda*
Phalaenopsis × *veitchiana*	*Phalaenopsis* × *veitchiana*	*Phalaenopsis* × *veitchiana*

(continued)

Table 10.1 (continued)

Sweet (1980)	Shim (1983)	Christenson (2001)
Section *Stauroglottis*	**Section *Stauroglottis***	**Section *Stauroglottis***
Phalaenopsis equestris	*Phalaenopsis equestris*	*Phalaenopsis equestris*
Phalaenopsis celebensis	*Phalaenopsis celebensis*	*Phalaenopsis celebensis*
Phalaenopsis lindenii	*Phalaenopsis lindenii*	*Phalaenopsis lindenii*
		Section *Deliciosae*
		Phalaenopsis chibae[a]
		Phalaenopsis deliciosa[a]
Genus *Doritis*	**Genus *Doritis***	**Section *Esmeralda***
Doritis pulcherrima	*Doritis pulcherrima*	*Phalaenopsis pulcherrima*
Doritis regnieriana	*Doritis regnieriana*	*Phalaenopsis regnieriana*
Doritis buyssoniana	*Doritis buyssoniana*	*Phalaenopsis buyssoniana*

[a]New species [b]Revised species

Table 10.2 The 12 commercialized *Phalaenopsis* cultivars studied (from Tsai et al. 2015)

Cultivars name	Floral color	Flower size in diameter (cm)
P. Sogo Yukidian 'V3'	White	12–13
P. Sogo Musadian	White	12–13
P. I-Hsin Diamond	White	12–13
P. Chainport Dorothy	White	12–13
P. Ruey Lih Beauty	Red	8–10
P. Shiuh-Dong Red Rose 'Fantasy Rose'	Red	8–10
P. Ruey-Lih Red Rose	Red	8–10
OX 1172	Red	8–10
P. Sogo Meili 'Sogo F1751'	Yellow	6–7.5
Sogo F3005	Yellow	6–7.5
P. Sogo Shito 'Sogo F2999'	Yellow	6–7.5
P. Sogo Sweet	Yellow	6–7.5

Deliciosae of subgenus *Phalaenopsis*. Therefore, the subgenus *Phalaenopsis* was not monophyletic based on the molecular data. In the subgenus *Phalaenopsis*, the nrITS data shows that section *Phalaenopsis* is monophyletic (Yukawa et al. 2005; Tsai et al. 2006) rather than based on the plastid matK, atpH-F, and trnD-E spacers (Padolina et al. 2005) and the plastid matK and trnK introns (Yukawa et al. 2005). According to nrITS data, the subgenus *Polychilos* is polyphyletic (Yukawa et al. 2005; Tsai et al. 2006) but not established on the plastid DNA (Padolina et al. 2005; Yukawa et al. 2005). In the *Polychilos* subgenus, only section *Polychilos* was shown to be monophyletic according to the internal transcribed spacers (ITS) data (Yukawa et al. 2005; Tsai et al. 2006), and section *Fuscatae* was shown to be monophyletic based on plastid DNA.

Furthermore, Tsai et al. (2009) examined the molecular phylogeny of *Phalaenopsis* Blume (Orchidaceae). The internal transcribed spacers

(ITS) of nuclear ribosomal DNA (nrDNA) and plastid DNA, with the *trn*L intron, the *trn*L-F spacer, and the *atp*B-*rbc*L spacer, were used to reconstruct the phylogeny of this genus. The monophyly of *Phalaenopsis* was sustained by the nrITS, plastid DNA, and combined DNA data sets. Therefore, the subgenus *Polychilos* of *Phalaenopsis* was monophyletic, and the species were separated into two subclades. Moreover, since the *Esmeralda* and *Deliciosae* seemed to separate from sections *Phalaenopsis* and *Stauroglottis,* the subgenus *Phalaenopsis* proved to be non-monophyletic, which agrees with the Ko et al. (2017). Meanwhile, the subgenera *Aphylae* and *Parishianae* were also revealed to be non-monophyletic according to the molecular data. Additionally, the monotypic species *P. lowii* of subgenus *Proboscidioides* created a clade with subgenus *Aphyllae* (Figs. 10.1 and 10.2).

10.4 Plastid *trn*L Intron Polymorphisms Among *Phalaenopsis* Species

Tsai et al. (2012) determined the *trn*L intron sequences of the plastid DNA of more than 95% *Phalaenopsis* living native species and detected 54 native *Phalaenopsis* species by using the plastid *trn*L intron sequence. The inheritance of the plastid genome of the three interspecific hybrids within *Phalaenopsis* species was constructed based on the examination of the *trn*L intron sequence. Moreover, the identification of the plastid genome type of various *Phalaenopsis* hybrids used native *trn*L sequences. It was found that most *Phalaenopsis* species have unique *trn*L intron sequences, which are caused by mutations, insertions/deletions, or both (Fig. 10.3). Molecular evidence indicated that the maternal inheritance of the plastid genome appears in the interspecific hybrids of *Phalaenopsis* species. This analysis revealed the plastid genome type of the hybrids; it could help to re-evaluate the genealogies since the plastid DNA is maternally inherited. The *trn*L intron sequences from three *Phalaenopsis* hybrids including *P.* Yungho Gelb Canary, *P.* Timothy Christopher, and *P.* Rainbow Chip were examined in order to re-evaluate their genealogies from the Sander's List of Orchid Hybrids. Nevertheless, the examination of *Phalaenopsis* hybrids resulted in no heterogeneous *trn*L intron sequences in any of them. The *trn*L intron sequence from *P.* Yungho Gelb Canary (accession number: FJ705059) was the same with *P. amboinensis,* suggesting that the plastid genome type for the hybrid was inherited from *P. amboinensis*. The result matched with the genealogy of *P.* Yungho Gelb Canary, which was reported in the Wildcatt Database (Fig. 10.4a). The plastid genome type of the *trnL* intron sequence from *P.* Timothy Christopher (accession number: FJ472585) showed that this hybrid was inherited from *P. stuartiana*, which was in agreement with the Wildcatt Database. It is suggested that the plastid genome type of *P.* Timothy Christopher was inherited from *P. equestris* (Moir 1995). Additionally, the genealogy of *P.* Timothy Christopher obtained from the Wildcatt Database showed that *P.* Cassandra was the maternal parent and *P. amabilis* was the paternal parent consistent with the registration by Sandrik in 1982. This examination showed that the registration for *P.* Cassandra was incorrect. In order to verify this result, the maternal parent *P.* Cassandra was further assessed. *P.* Cassandra was obtained via hybridization between *P. equestris* as the maternal parent and *P. stuartiana* as the paternal parent according to the reports. The genealogy was recorded by Veitch in 1896 (see Moir 1995) (Fig. 10.4b). However, the plastid genome type of the *P.* Cassandra was resolute based on its *trnL* intron sequence (accession number: JQ613334). The results indicated that the plastid genome type for *P.* Cassandra was inherited from *P. stuartiana*, thus *P. stuartiana* is the maternal parent, not *P. equestris*. Therefore, it was concluded that the genealogy of *P.* Cassandra registration was not correct. The inaccurate genealogy of *P.* Cassandra could imitate the genealogy of the afterward generation hybrid, *P.* Timothy Christopher. The *trn*L

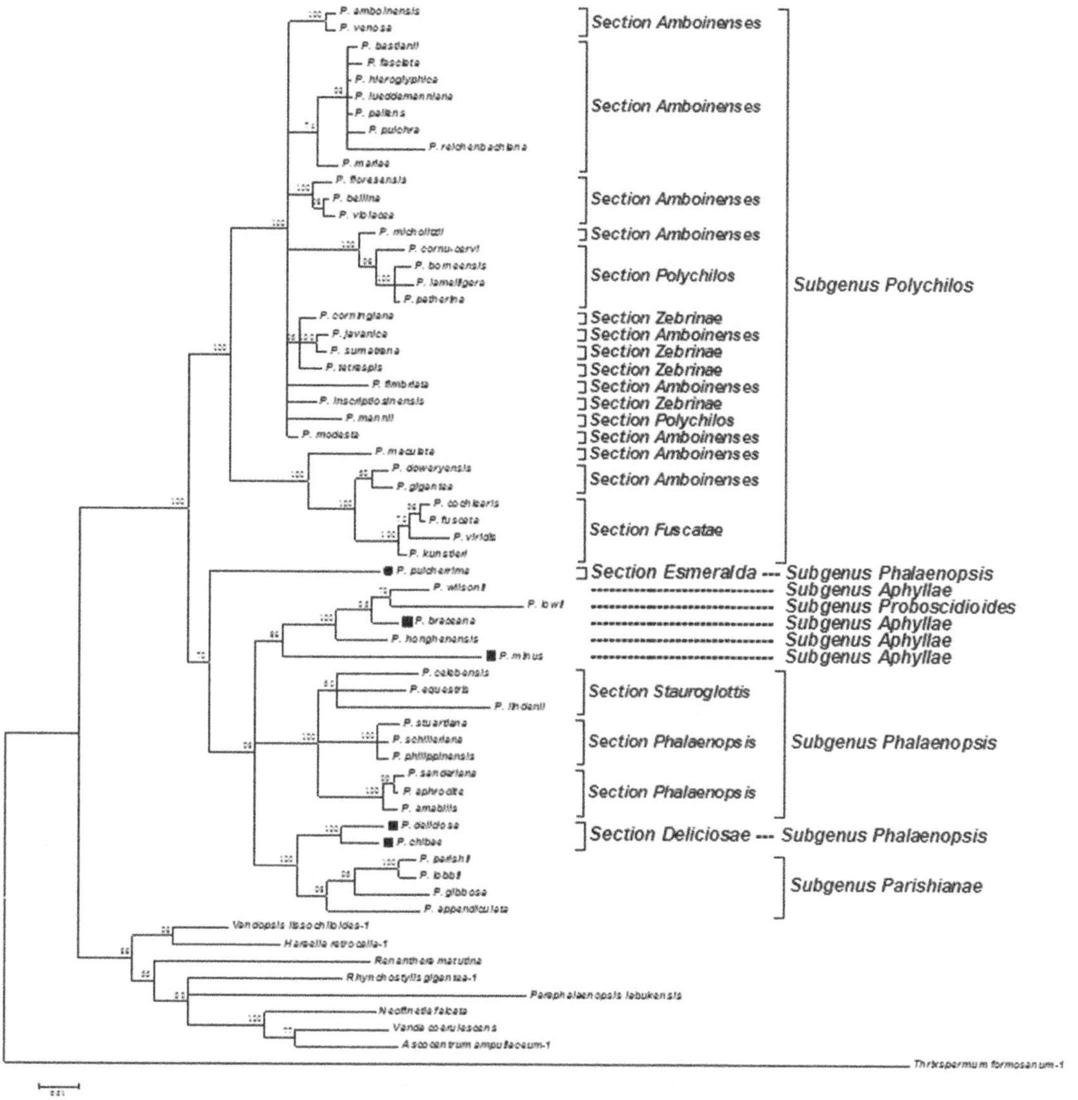

Fig. 10.1 The most parsimonious tree results obtained from the analysis of the combined data matrix (the nuclear ribosomal ITS, and plastid *trn*L intron, *trn*L-F spacer, and *atp*B-*rbc*L spacer) of 52 *Phalaenopsis* and 9 outgroup species. Bootstrap values >50% were revealed over each branch. Solid circle (●) shows that the species was traditionally regarded as the genus *Doritis*. Solid squares (■) indicated that the species was usually regarded as the genus *Kingidium*. (Redrawn from Tsai et al. 2010a)

intron sequence (accession number: FJ472586) of another hybrid *P.* Rainbow Chip indicated that the plastid genome type of the hybrid was inherited from *P. stuartiana*. However, it does not agree with the genealogy of the hybrid from the Wildcatt Database. It is suggested that the plastid genome type could be inherited from *P. equestris* (Moir 1995) (Fig. 10.4c). Additionally, *P.* Cassandra was registered as the maternal parent of *P.* Rainbow Chip from the analysis of the genealogy of this hybrid. Consequently, the incorrect genealogy of *P.* Rainbow Chip was due to the incorrect recording of *P.* Cassandra and the prior inspection of *P.* Timothy Christopher.

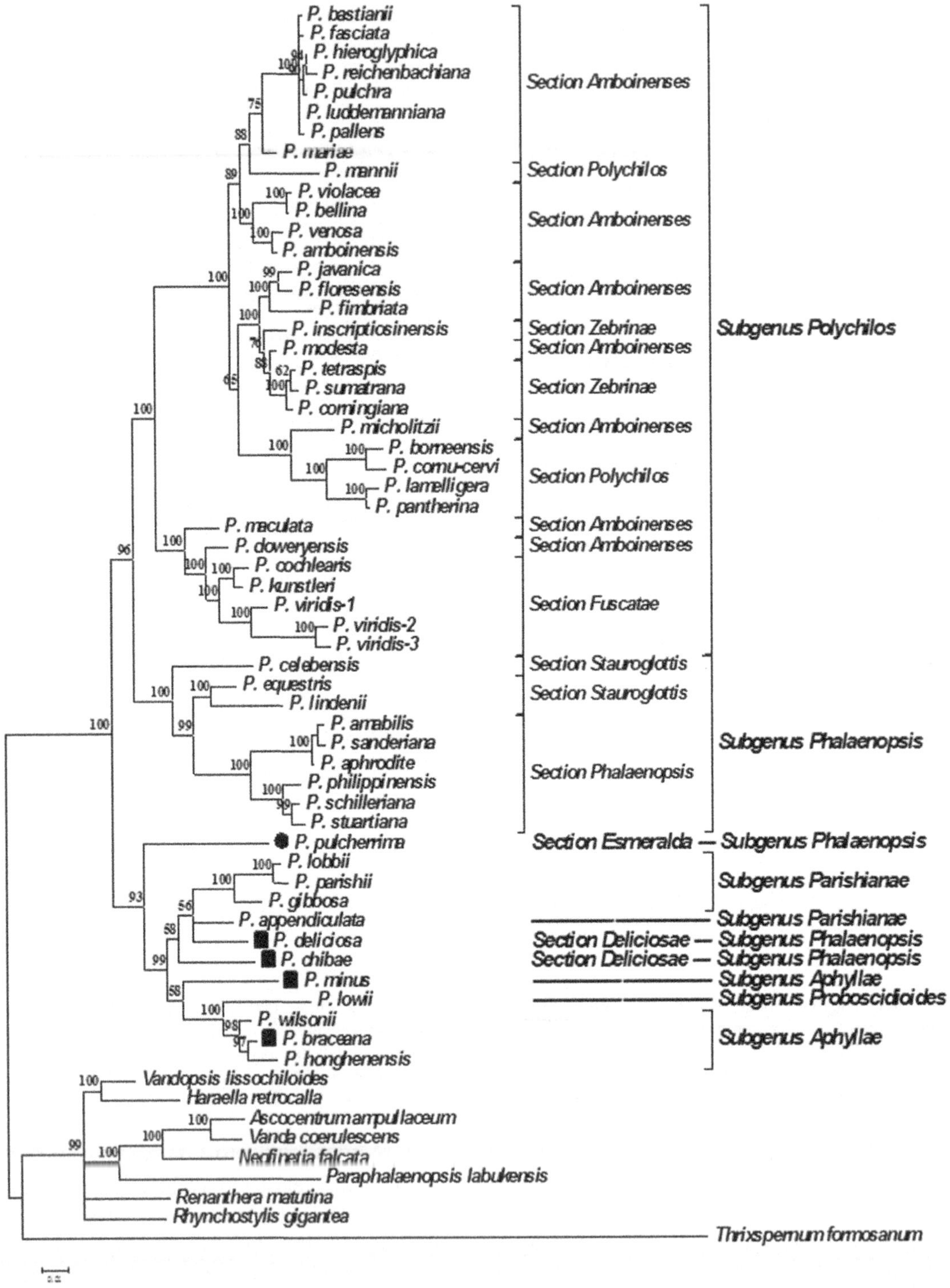

Fig. 10.2 The Bayesian inference tree was derived from the analysis of the combined data matrix (the nuclear ribosomal ITS, plastid trnL intron, trnL-F spacer, and atpB-rbcL spacer) based on 52 *Phalaenopsis* and 9 outgroup species. Posterior probabilities >50% were revealed over each branch. Solid circle (●) demonstrated that the species was traditionally regarded as the genus Doritis. Solid squares (■) indicated that the species was traditionally regarded as the genus *Kingidium*. (Redrawn from Tsai et al. 2010a)

```
P._amabilis         G-----CCGG GCGATCGAAT GAT------- ---------- -------GAA ATTTCGCTTT CAAT------ ---AATAATA
P._aphrodite        .-----..A. -......... ...------- ---------- -------... .......... ....------ ---.......
P._sanderiana       .-----.AA. -......... ...------- ---------- -------... .......... ....------ ---.......
P._intermedia       .-----..A. -......... ...------- ---------- -------... .......... ....------ ---.......
P._philippinensis   .GAAAA.... -..------- ...------- ---------- -------... .......... ..T.TTTCTT TCT.TA.TA.
P._schilleriana     .GAAAA.... -..------- ...------- ---------- -------... .......... ..T.TTTCTT TCT.TA.TA.
P._stuartiana       .GA---...A -..------- ...------- ---------- -------... .......... ..T.TTTCTT TCT.TA.TA.
P._celebensis       .-----.... -..------- ...------- ---------- -------... ....A.A... A.T.------ ---.......
P._equestris        .-----.... -..------- .--------- ---------- -------... ......---- ..T.------ ---.......
P._lindenii         .-----.... -..------- ...TTAGATC TATATATTTA GATATAT... ....A..... ..T.------ ---...C..T
P._deliciosa        .-----...A -.A------- ...------- ---------- -------... ....AT.... ..T.------ ---.......
P._chibae           A-----.... -AA------- ...------- ---------- -------... ....A..... ..T.------ ---C......
P._pulcherrima      .TAAAAA... -..------- A..------- ---------- ---------- ----...... .--------- ---..A..C.

P._amabilis         ---------- ---------- TCTAA----- --AATATGAG TAATAT---- -----G-ACG GC
P._aphrodite        ---------- ---------- .....----- --........ ......---- -----.-... ..
P._sanderiana       ---------- ---------- .....----- --........ ......---- -----.-... ..
P._intermedia       ---------- ---------- .....----- --........ ....C.---- -----.-... ..
P._philippinensis   ---------- -------TAT .....AATAT GAGTA..ATA ......---- -----.-.A. T.
P._schilleriana     ---------- -------TAT .....AATAT GAGTA..ATA ......---- -----.-.A. T.
P._stuartiana       ---------- -------TAT .....AATAT GAGTA..ATA ......---- -----.-.A. T.
P._celebensis       TAATATGAGT ATTATATTAT .....----- --........ ......---- -----.-.AT .T
P._equestris        ---------- ---ATATTAT .--------- ---------- ----..---- -----.-.A. ..
P._lindenii         C--TATTAAT ATTATATTAT .....----- --........ ......TTTA ATTAG.-TA. ..
P._deliciosa        ---------- ---------- .....----- --........ ......---- -----T-.A. ..
P._chibae           ---------- ---------- G....----- --........ ......---- -----T-.A. ..
P._pulcherrima      ---------- -------TAT .--------- ---------- ----.G---- -----TG.A. ..
```

Fig. 10.3 Polymorphic sites of the aligned sequences of species of the subgenus *Phalaenopsis*. Dots (…) show identical nucleotides, and gaps (—) show insertions or deletions. (Redrawn from Tsai et al. 2012)

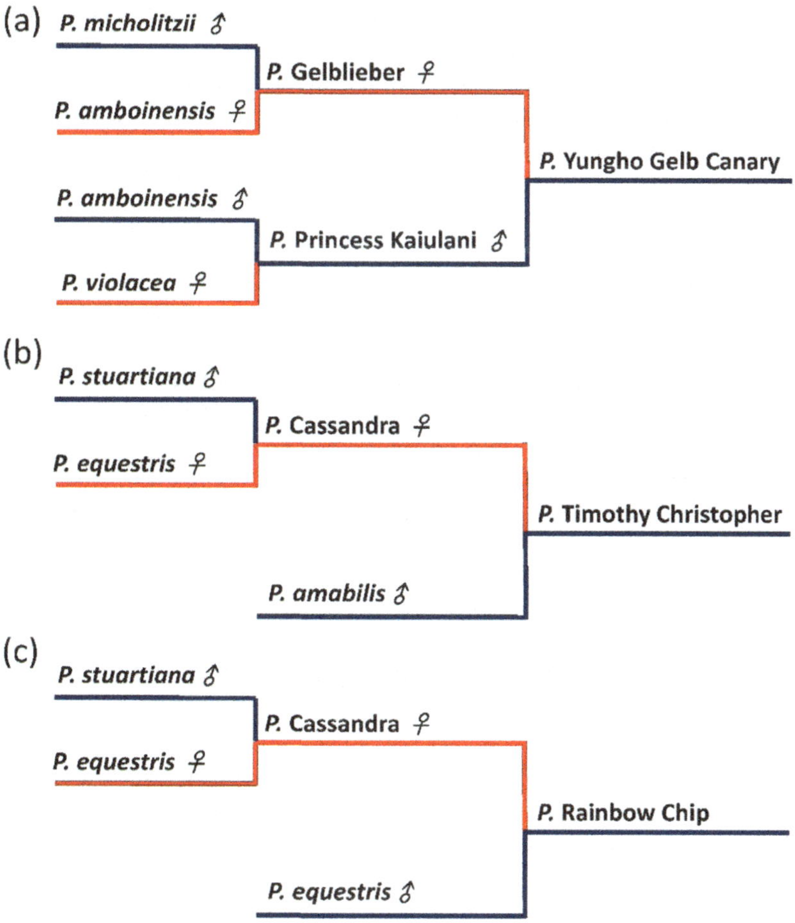

Fig. 10.4 Genealogies of *Phalaenopsis* Yungho Gelb Canary (**a**), *P.* Timothy Christopher (**b**), and *P.* Rainbow Chip (**c**). These genealogies were redone from the Wildcatt Database. (Redrawn from Tsai et al. 2012)

10.5 RNA-Seq SSRs of Cultivars

To improve the commercial cultivars of Phalaenopsis orchids, the molecular markers could be an essential equipment. Lately, the amplified fragment length polymorphism (AFLP) (Kashkush et al. 2001), inter-simple sequence repeat (ISSR) (Eiadthong et al. 1999), and microsatellite markers (simple sequence repeats, SSRs) (Tsai et al. 2013) of DNA markers have been used to characterize cultivars, in which the significant recurrent motifs of microsatellites are disposed to mutation due to slipped-strand mispairing (Levinson et al. 1987). However, the advantages of SSR markers are their high dependability; high polymorphism, codominance, and transferability in related species (Rafalski et al. 1996); and high frequency in genome. Moreover, microsatellites are cycle repeats of short (2–6 bp) DNA sequences. The microsatellite markers have been isolated from tropical fruit trees and categorized, such as mango (Chiang et al. 2012), Indian Jujube (Chiou et al. 2012), and guava (Rai et al. 2013). Thus, SSR markers have been confirmed to be useful for investigating genetic relationships in related plant species and subpopulations of individual species (Bowcock et al. 1994).

10.6 RNA-Seq SSRs of the *Phalaenopsis*

Young (2004) assessed the DNA fingerprinting of 89 accessions of *Phalaenopsis amabilis* constructed on microsatellite DNA (simple sequence repeats, SSRs). Three SSR loci were duplicated and assessed from *P. amabilis* accessions. These loci are useful molecular markers to detect the intraspecific variation of *Phalaenopsis*. Moreover, 42 (Han 2005) and 261 EST-SSR loci (Zhang et al. 2013) were obtained from EST-SSRs, which were individually established from the *Phalaenopsis* EST database. The large-scale BAC end sequencing discovered 950 potential SSRs in *Phalaenopsis equestris* (Hsu et al. 2011). Therefore, deep sequencing skills give the option of generating large number of SSR markers in a fast and at a lower cost method than library-based methods (Abdelkrim et al. 2009; Santana et al. 2009; Cavagnaro et al. 2010; Csencsics et al. 2010; Zhu et al. 2012).

Furthermore, the genus *Phalaenopsis* was examined by Tsai et al. (2015) using the fast sequence tag (EST)-simple sequence repeat (SSR) markers. SSR appears in the exon region, such as the 5' untranslated region (UTR), coding region (CDS), and 3'UTR, on average every 20.22 kb in *P. aphrodite* subsp. *formosana*. Tsai et al. (2015) completed de novo transcriptome deep sequencing of *P. aphrodite* subsp. *formosana* to examine EST-SSR. Ten polymorphic and transferable SSR loci across the 22 native taxa could be obtained. Subsequently, the identification of 12 commercial *Phalaenopsis* cultivars was done in the EST-SSR study, containing white, red, and yellow floral color groups (Table 10.3). The morphological types of the same floral color from the plant materials were very similar. It was not easy to distinguish them based on either vegetative or reproductive characters (such as floral color, size, and morphology). Three validated polymorphic and transferable primer pairs for EST-SSR were selected to distinguish 12 commercial *Phalaenopsis* cultivars. In the white floral color group (Fig. 10.6a–d), each of cultivars could be recognized from SSR loci Pap-3222 and Pap-4282 (Fig. 10.5a). Every cultivars could be recognized in red floral color group (Fig. 10.6e–h) and yellow floral color group (Fig. 10.6i–l) by using SSR loci Pap-4825 (Fig. 10.5b) and Pap-3222 (Fig. 10.5c). Thus, all 12 commercial *Phalaenopsis* cultivars could be distinguished by using the aforementioned three EST-SSR markers. Moreover, over two bands might deviate for an individual in the analysis of 12 commercial *Phalaenopsis* cultivars (Fig. 10.5). The ploidy of commercial *Phalaenopsis* cultivars is generally triploid or tetraploid (Grabherr et al. 2011). Additionally, 12 *Phalaenopsis* cultivars could be distinguished according to the analysis of three SSR loci (Fig. 10.5). Hence, the EST-SSR markers could be used to classify *Phalaenopsis*

Table 10.3 Information on geographic distribution, species code, and voucher specimens of the genus *Phalaenopsis* used in this study (from Ko et al. 2017)

Classification	Geographical distribution	Source[a]
Subgenus *Proboscidioides* (Rolfe) E. A. Christ.		
P. lowii Rchb.f.	Myanmar, and adjacent western Thailand	KDAIS-KC88
Subgenus *Aphyllae* (Sweet) E. A. Christ.		
P. minus (Seidenf.) E. A. Christ.	Endemic to Thailand	KDAIS-KC227
P. braceana (J. D. Hook.) E. A. Christ.	Bhutan and China	KDAIS-KC289
Subgenus *Parishianae* (Sweet) E. A. Christ.		
P. parishii Rchb.f.	Eastern Himalayas, India, Myanmar, and Thailand	KDAIS-KC316
Subgenus *Polychilos* (Breda) E. A. Christ.		
P. mannii Rchb.f.	Northeast India, Nepal, and China to Vietnam	KDAIS-KC22
P. cornu-cervi (Breda) Bl. and Rchb.f.	Northeast India and the Nicobar Islands to Java and Borneo	KDAIS-KC23
P. kunstleri J. D. Hook.	Myanmar and Malay Peninsula	KDAIS KC-139
P. pulchra (Rchb.f.) Sweet	Endemic to the Philippines (Luzon and Leyte)	KDAIS-KC17
P. violacea Witte	Indonesia (Sumatra) and Malaysia (Malay Peninsula)	KDAIS-KC153
P. micholitzii Rolfe	Philippines (Mindanao)	KDAIS-KC382
P. maculata Rchb.f.	Malaysia (Pahang), East Malaysia (Sabah and Sarawak), and Indonesia (Kalimantan Timur)	KDAIS-KC49
P. amboinensis J. J. Sm.	Indonesia (Molucca Archipelago and Sulawesi)	KDAIS-KC157
P. inscriptiosinensis Fowlie	Endemic to Indonesia (Sumatra)	KDAIS-KC298
P. corningiana Rchb.f.	Borneo (Sarawak and elsewhere on the island)	KDAIS-KC346
Subgenus *Phalaenopsis*		
P. amabilis (L.) Blume	Widespread from Sumatra and Java to the southern Philippines, east to New Guinea and Queensland, Australia	KDAIS-KC23
P. aphrodite Rchb.f.	Northern Philippines and southeastern Taiwan	KDAIS-KC96
P. schilleriana Rchb.f.	Endemic to the Philippines	KDAIS-KC429
P. chibae Yukawa	Yukawa endemic to Vietnam	KDAIS-KC488
P. pulcherrima (Lindl.) J. J. Sm.	Widespread from northeast India and southern China throughout Indochina to Malaysia (Malay Peninsula), Indonesia (Sumatra), and East Malaysia (Sabah)	KDAIS-KC256

(continued)

Table 10.3 (continued)

Classification	Geographical distribution	Source[a]
P. equestris (Schauer) Rchb.f.	Philippines and Taiwan	KDAIS-KC203
P. lindenii Loher	Endemic to the Philippines	KDAIS-KC119

[a]Plant materials were cultivated at the Kaohsiung District Agricultural Improvement Station, Taiwan and voucher specimens were deposited at the herbarium of the National Museum of Natural Science, Taiwan

cultivars. The deep sequencing is definitely a fast approach for obtaining the sequences needed to determine SSRs and for designing exact primers to get suitable SSR markers.

In addition, the EST-SSR markers generally have greater amplification efficiency and are easier to transfer between species than SSR markers resultant from genomic non-coding regions (Gupta et al. 2003; Varshney et al. 2005; Wang et al. 2010; Garcia et al. 2011). Therefore, the transferability and polymorphisms of SSR loci cloned from EST-SSRs need to be validated in order to develop the universal SSR markers for all commercial *Phalaenopsis* cultivars. Transcrip-

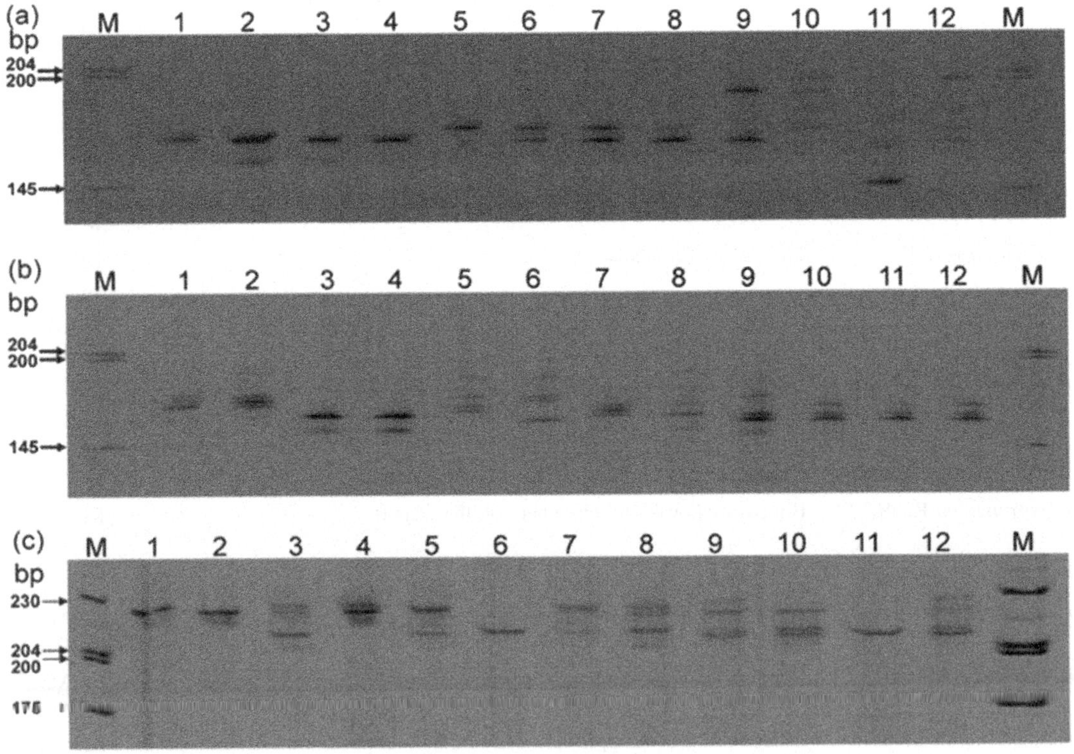

Fig. 10.5 The polymorphism of 12 *Phalaenopsis* cultivars by SSR–PCR analysis. The polymorphism of 12 *Phalaenopsis* cultivars at **a** Pap-3222 SSR, **b** Pap-4825 SSR, and **c** Pap-4282 SSR loci. Lanes 1–12 represent 12 *Phalaenopsis* cultivars/lines listed in Table 10.4. Lanes 1–4 represent four similar commercial cultivars with white floral color; Lanes 5–8 represent four related commercial cultivars with yellow floral color; Lanes 9–12 represent four similar commercial cultivars with red floral color. (From Tsai et al. 2015)

Fig. 10.6 The flower of 12 different *Phalaenopsis* cultivars. Images **a–l** represent samples 1–12 shown in Table 10.2. (From Tsai et al. 2015)

tome analysis constructed on deep sequencing is an useful approach for the development of SSR loci in non-model species. A large number of EST-SSR loci could be isolated after the preliminary validation; approximately 33.33% of EST-SSR loci are general markers across *Phalaenopsis* breeding germplasm.

10.7 Transferable Microsatellite Markers of *Phalaenopsis*

Ko et al. (2017) used more microsatellite loci and more extensive species testing than earlier studies (Sukma 2011; Tsai et al. 2015) to enhance the ability to distinguish the *Phalaenopsis* genus. The genome size of *P. aphrodite* is small. It is appropriate for the development of microsatellite markers (Hsiao et al. 2011) and in the diploid genome size of approximately 2.81 pg (Chen et al. 2013). The modified magnetic bead enrichment method was used to develop the convenient microsatellite markers from *P. aphrodite* subsp. *formosana*. The molecular identification systems could be established based on these transferable markers to access the hybridization and introgression among the species of genus *Phalaenopsis* in the future. Moreover, Ko et al. (2017) examined 21 species of *Phalaenopsis,* including five subgenera (Table 10.4), which matched with the general taxonomy and nomenclature (Christenson 2001). A total of 146 repeatable amplicons with variation in length were selected using 28 microsatellite primer pairs (Table 10.4) from 21 species (Table 10.4). Subsequently, the 28 microsatellite primers from *P. aphrodite* subsp. *formosana* were used to conduct the cross-species amplification test for the other 20 species. The species of *P. amabilis* (L.) Blume, *P. schilleriana* Rchb.f, *P. chibae*, *P. equestris* (Schauer) Rchb.f., and *P. lindenii* Loher have high transferable loci, which are all categorized under the genus *Phalaenopsis,* including *P. aphrodite* subsp. *formosana*. Therefore, these newly developed microsatellite primers could be used to develop a standard molecular identification operating system in *Phalaenopsis* due to the high transferability property of the species of the subgenus *Phalaenopsis*.

Furthermore, a model-based Bayesian clustering algorithm was conducted in STRUCTURE 2.3.4 for genetically delimiting 21 species of the genus *Phalaenopsis*. At the first optimal

Table 10.4 Characteristics of the 28 polymorphic microsatellite primers derived from *Phalaenopsis aphrodite* subsp. *formosana* (from Ko et al. 2017)

Locus	Primer sequences (5'–3')		Repeat motif	Allelic size (bp)	Annealing temperature (°C)	Number of alleles	PIC
PA5-1	F:	TCCCATTATCACTCCCTCAC	$(TC)_{14}$	140–164	59	4	0.67
	R:	GGTTAGA ATATAGGGAGAG					
PA5-2	F:	CTCTCTTTCCTTCTCACCTC	$(TC)_{10}$	98–104	58	4	0.61
	R:	AAGATAGAGGGAGAGAGTGG					
PA7	F:	CTCTGCTTCTCACCTTTCAC	$(TC)_{12}$	116–264	56	3	0.55
	R:	GGACAGAAAGTGAGAGAGAG					
PA10	F:	TCTTCAGTCCCTCACTCATC	$(CT)_{14}$	132–152	58	7	0.75
	R:	ACAAAGCGGTGGAGAATATG					
PA11	F:	ATCTATTGCTCTTTGTCCTC	$(CT)_{42}$	214–216	55	2	0.38
	R:	TAGCAAAGAGATGCTGAAGG					
PA14	F:	TTTTCACTCTCCCTCCATCC	$(CT)_{21}$	182–186	52	3	0.55
	R:	GATGTAGAGAATGAGGGAGC					
PA15	F:	TCTCCTACTCCCTCTATCTCA	$(CT)_{25}$	306–310	56	3	0.55
	R:	CTTGAAAGGCAGAGAGATAG					
PA19	F:	TCTCCCTATATCTCTGCATC	$(CTCC)_4$	154–158	51	3	0.58
	R:	TGGAAAGAGAAAGGTTCAGG					
PA21	F:	TCTCTCACTTTGTCACTCGC	$(CT)_{14}$	134–146	57	6	0.77
	R:	AAAGGGAAGTAGGGAAGGAG					
PA24	F:	TTGATCTCTCTGGCACCCAC	$(TC)_{36}$	216–224	55	3	0.59
	R:	AAGAGAGAGTTAGTTGGAGAT					
PA25-1	F:	ACCCACTTTCTCCTATCTCC	$(CT)_{20}$	176–202	58	5	0.64
	R:	GATGAAAGAGAGTGAGAGCG					
PA25-2	F:	TCTCCCTCTCTTTACCACTC	$(CT)_{12}$	92–267	58	6	0.73
	R:	GTGAGAGAGATAGAGTGAGC					
PA32-1	F:	CTCTTCCTGCTTTTCCTAGG	$(CT)_{25}$	148–222	57	8	0.83
	R:	AAGAGGGTGTGAGGAAGAGG					
PA32-2	F:	TCTCTCACTACTCTATCTTG	$(CT)_{18}$	140–152	54	5	0.73
	R:	GAGAAGATAGAAAGAGTGAG					
TPA36	F:	CTCCACTTTATCTCTCTACC	$(TC)_{39}$	220–250	55	4	0.67
	R:	ATTGAGCGAGATAAAACTAG					
PA37	F:	TTTACCTCTTTTGCTAGCTC	$(TC)_{23}$	226–234	50	4	0.67
	R:	AAGAGAAAGGGAAGGAGAGC					
PA38	F:	CTCTCTCACTCTATTACTCC	$(CT)_{32}$	224–384	54	9	0.84
	R:	AGCTAGATAGAGGGAGAAAG					
PA40	F:	GAGCAACATTCACTAGAGAG	$(CA)_{14}$	258–320	56	6	0.79
	R:	CTGGCAAAGCTTTGAGAAGG					

(continued)

Table 10.4 (continued)

Locus	Primer sequences (5'–3')		Repeat motif	Allelic size (bp)	Annealing temperature (°C)	Number of alleles	PIC
PA41	F:	GAGGAGAAATAATGATTCCG	$(AG)_{12}$	138–140	50	2	0.38
	R:	AGACACTCTCACACACTTTC					
PA63	F:	TTCATTCCATCTACCCCATC	$(CT)_8$	130–136	55	3	0.59
	R:	GATAGAAAGACTAGAGTAGG					
PA64	F:	CTCTCCTTTTTCTTATCTTTCAC	$(CT)_{94}$	248–296	55	3	0.55
	R:	TAGAGAGATAGAGGGCAAGC					
PA74-1	F:	AATGACCTCTCTGCTCTCTC	$(TC)_{28}$	172–306	50	3	0.59
	R:	GCAAGAGAAGTTGTGGGATGG					
PA74-2	F:	CATCCCACAACTTCTCTTGC	$(CT)_{13}$	122–134	55	5	0.58
	R:	AGTGCTCAAGCGAGTTAGAGAC					
PA83-1	F:	CCCTCTTTCTCTCATTGTCC	$(TC)_9$	190–198	54	2	0.35
	R:	GGGACAGAGTGCATAAGATG					
PA83-2	F:	CCTTATCTCTTCTCTCTACC	$(TC)_{36}$	168–210	50	12	0.87
	R:	AGAAAGGAAGGGTAGGAGAG					
PA100-1	F:	TCCCTCTATTTTAGACACCC	$(TC)_{11}$	132–136	52	2	0.35
	R:	GGAGAAAGAGCAAGACAGTG					
PA100-2	F:	TCTCCATCCGTTAGCCTCTC	$(CT)_{16}$	128–136	59	5	0.73
	R:	GGGTAGGCAGAGAGAGTGAT					
PA101	F:	CCCACTCACACTCTATCTTC	$(TC)_{11}$	126–138	55	7	0.63
	R:	AGGGTCAAACAGAATGAAGG					
					Average	4.57	0.63

clustering number K = 2, most species of the subgenus *Phalaenopsis* were allocated to the similar cluster with a great percentage of Component 1 (pink segment in Fig. 10.7a) except for *P. pulcherrima*, which is genetically allocated to section Esmeralda (subgenus Polychilos). Moreover, the subgenera of *Proboscidioides*, *Aphyllae*, *Parishianae*, and Polychilos were transferred to the cluster with a great percentage of Component 2 (Fig. 10.7a) except for *P. kunstleri* that explicated an admixture genetic composition. Furthermore, at the second optimal clustering number K = 4, sections *Deliciosae* and *Esmeralda* could be separated into different clusters, which are gathered when K = 4, and this result corresponds with that of Tsai et al. (2010a). Additionally, two sections of *Phalaenopsis* and *Stauroglottis* from subgenus *Phalaenopsis* were gathered with great genetic match (Fig. 10.7b). The section of subgenus *Polychilos* was genetically allocated to the subgenus *Phalaenopsis* cluster constructed on section *Fuscatae* of subgenus *Polychilos* (Fig. 10.7a, b). The assignment test of Bayesian clustering analysis showed a close result similar to the molecular phylogeny patterns that were characterized by Tsai et al. (2005). Bayesian clustering analysis constructed on EST-SSR loci was not enough to distinguish among subgenus or sections within subgenus (Tsai et al. 2015). Therefore, the lately developed genomic microsatellite loci have greater resolution ability compared with the EST-SSR loci when studying on native moth orchids.

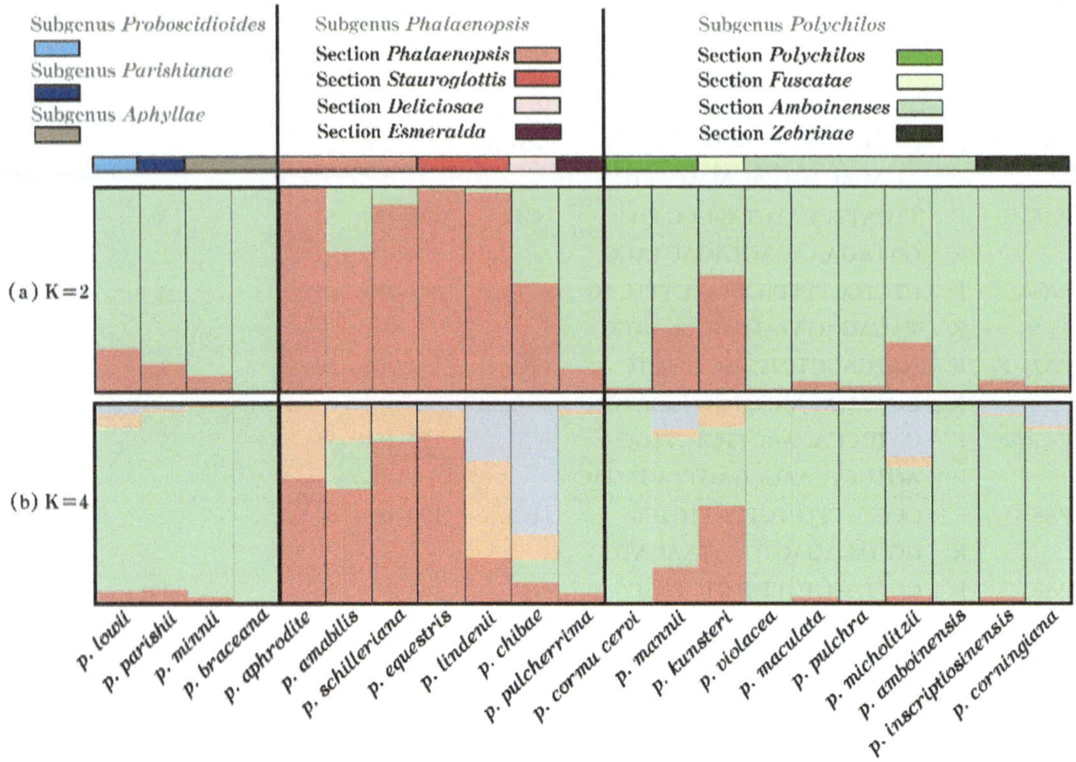

Fig. 10.7 Genotypic group structure of the genus *Phalaenopsis*, which was used in this study in K = 2 (**a**) and 4 (**b**). (Redrawn from Ko et al. 2017)

References

Abdelkrim J, Robertson BC, Stanton JAL, Gemmell NJ (2009) Fast, cost-effective development of species-specific microsatellite markers by genomic sequencing. Biotechniques 46:185–192

Bowcock AM, Ruiz-Linares A, Tomfohrde J, Minch E, Kidd JR, Cavalli-Sforza LL (1994) High resolution of human evolutionary trees with polymorphic microsatellites. Nature 368:455–457

Cai J, Liu X, Vanneste K, Proost S, Tsai WC, Liu KW et al (2015) The genome sequence of the orchid *Phalaenopsis equestris*. Nat Genet 47:65–72

Chang CC, Lin HC, Lin IP, Chow TY, Chen HH, Chen WH (2006) The chloroplast genome of *Phalaenopsis aphrodite* (Orchidaceae): Comparative analysis of evolutionary rate with that of grasses and its phylogenetic implications. Mol Biol Evol 23:279–291

Cavagnaro PF, Senalik DA, Yang L, Simon PW, Harkins TT, Kodira CD, Huang S, Weng Y (2010) Genome-wide characterization of simple sequence repeats in cucumber (*Cucumis sativus* L.). BMC Genom 11:569–586

Chen WH, Chen HH (2007) Orchid Biotechnology I. World Sci Pub, Singapore

Chen WH, Chen HH (2011) Orchid Biotechnology II. World Sci Pub, Singapore

Chen WH, Kao YL, Tang CY, Tsai CC, Lin TY (2013) Estimating nuclear DNA content within 50 species of the genus *Phalaenopsis* Blume (Orchidaceae). Sci Hortic 161:70–75

Chiang YC, Tsai CM, Chen YKH, Lee SR, Chen CH, Lin YS, Tsai CC (2012) Development and characterization of 20 new polymorphic microsatellite markers from *Mangifera indica* L. (Anacardiaceae). Am J Bot 99(5):e117–e119

Chiou CY, Chiang YC, Chen CH, Yen CR, Lee SR, Lin YS, Tsai CC (2012) Development and characterization of 38 polymorphic microsatellite markers from an economical fruit tree, the Indian jujube. Am J Bot 99(5):e199–e202

Christenson EA (2001) *Phalaenopsis*: a Monograph. Timber Press, Portland, Oregon

Csencsics D, Brodbeck S, Holderegger R (2010) Cost-eVective, species-specific microsatellite development for the endangered dwarf bulrush (*Typha minima*) using next-generation sequencing technology. J Hered 101:789–793

Dressler RL (1993) Phylogeny and Classification of the Orchid Family. Cambridge Univ Press, Cambridge

Eiadthong W, Yonemori K, Sugiura A, Utsunomiya N, Subhadrabandhu S (1999) Identification of mango cultivars of Thailand and evaluation of their genetic variation using the amplified fragments by simple sequence repeat-(SSR-) anchored primers. Sci Hortic 82:57–66

Garcia RAV, Rangel PN, Brondani C, Martins WS, Melo LC, Carneiro MS (2011) The characterization of a new set of EST-derived simple sequence repeat (SSR) markers as a resource for the genetic analysis of *Phaseolus vulgaris*. BMC Genet 12:41–54

Grabherr MG, Haas BJ, Yassour M, Levin JZ, Thompson DA, Amit I (2011) Full-length transcriptome assembly from RNA-Seq data without a reference genome. Nat Biotechnol 29:644–652

Gupta PK, Rustgi S, Sharma S, Singh R, Kumar N, Balyan HS (2003) Transferable EST-SSR markers for the study of polymorphism and genetic diversity in bread wheat. Mol Genet Genom 270:315–323

Han SY (2005) Molecular cloning and characterization of cDNA-SSRs in *Phalaenopsis*. Master Thesis, Department of Biology, National Cheng Kung University, Tainan, Taiwan. (Chinese with English abstract)

Hsiao YY, Pan ZJ, Hsu CC, Yang YP, Hsu YC, Chuang YC, Shih HH, Chen WH, Tsai WC, Chen HH (2011) Research on orchid biology and biotechnology. Plant Cell Physiol 52(9):467–1486

Hsu CC, Chung YL, Chen TC, Lee YL, Kuo YT, Tsai WC (2011) An overview of the *Phalaenopsis* orchid genome through BAC end sequence analysis. BMC Plant Biol 11:3–13. https://doi.org/10.1186/1471-2229-11-3 PMID: 21208460

Kashkush K, Fang J, Tomer E, Hillel J, Lavi U (2001) Cultivar identification and genetic map of mango (*Mangifera indica*). Euphytica 122:129–136

Ko YZ, Shih HC, Tsai CC, Ho HH, Liao PC, Chiang YC (2017) Screening transferable microsatellite markers across genus *Phalaenopsis* (Orchidaceae). Bot Stud. 58(1):48

Lai JM, Tsai CC, Yen CR, Ko YZ, Chen SR, Weng IS, Lin YS, Chiang YC (2015) Molecular characterization of twenty polymorphic microsatellite markers in the polyploid fruit tree species *Syzygium samarangense* (Myrtaceae). Genet Mol Res 14(4):13013–13021

Levinson G, Gutman GA (1987) Slipped-strand mispairing: a major mechanism for DNA sequence evolution. Mol Biol Evol 4:203–221

Moir WWG (1995) An Orchid Database. Wildcatt Database Co., USA

Orchid Biology: Reviews and Perspectives. Cornell University Press, Ithaca, New York

Padolina J, Linder CR, Simpson BB (2005) A phylogeny of *Phalaenopsis* using multiple chloroplast markers. Selbyana 26:23–27

Powell W, Machray GC, Provan J (1996) Polymorphism revealed by simple sequence repeats. Trends Plant Sci 1(7):215–222

Rafalski JA, Vogel JM, Morgante M, Powell W, Andre C, Tingey SV (1996) Generating and using DNA markers in plants. In: Nonmammalian Genome Analysis. A Practical Guide (Birren, B. and Lai, E., eds.). Academic Press, San Diego, pp. 75–134.

Rai D, Lee BK, Dalman C, Golding J, Lewis G, Magnusson C (2013) Parental depression, maternal antidepressant use during pregnancy, and risk of autism spectrum disorders: population based case-control study. BMJ (Clinical research ed.) 346:f2059

Santana Q, Coetzee M, Steenkamp E, Mlonyeni O, Hammond G, Wingfield M, Wingfield B (2009) Microsatellite discovery by deep sequencing of enriched genomic libraries. Biotechniques 46 (3):217–223

Shim PS, Fowlie JA (1983) A new species of *Phalaenopsis* from Sulawesi (Celebes) formerly confused with *Phalaenopsis psilantha* Schltr., *Phalaenopsis venosa* Shim and Fowl. Orchid Dig 47:124–128

Sukma D (2011) Development of sequence-based microsatellite marker for *Phalaenopsis* orchid. Hayati J Biosci 18:71–76

Sweet HR (1980) The Genus *Phalaenopsis*. The Orchid Digest, Pomona, p 128p

Tanaka R, Kamemoto H (1984) Chromosomes in orchids: counting numbers. In: Arditti J (ed) Orchid Biology, Reviews and Perspectives, vol III. Cornell Univ Press, Ithaca, pp 323–412

Tang CY, Chen WH (2007) Breeding and development of new varieties in *Phalaenopsis*. In: Chen WH, Chen HH (eds) Orchid Biotechnology I. World Sci Pub, Singapore, pp 1–22

Topik H, Yukawa T, Ito M (2005) Molecular phylogenetics of subtribe *Aeridinae* (Orchidaceae): insights from plastid matK and nuclear ribosomal ITS sequences. J Plant Res 118:271–284

Tsai CC, Chen YKH, Chen CH, Weng IS, Tsai CM, Lee SR, Lin YS, Chiang YC (2013) Cultivar identification and genetic relationship of mango (*Mangifera indica*) in Taiwan using 37 SSR markers. Sci Hortic 164:196–201

Tsai CC, Chiang YC, Huang SC, Chen CH, Chou CH (2010a) Molecular phylogeny of *Phalaenopsis* Blume (Orchidaceae) based on the plastid and nuclear DNA. Plant Syst Evol 288:77–98

Tsai CC, Huang SC, Chou CH (2005) Molecular phylogeny of *Phalaenopsis* Blume (Orchidaceae) based on the internal transcribed spacer of the nuclear ribosomal DNA. Plant Syst Evol 256:1–16

Tsai CC, Huang SC, Chou CH (2009) Phylogenetics, biogeography, and evolutionary trends of the *Phalaenopsis sumatrana* complex inferred from nuclear DNA and chloroplast DNA. Biochem Syst Ecol 37:633–639

Tsai CC, Huang SC, Huang PL, Chen FY, Su YT, Chou CH (2006) Molecular evidences of a natural hybrid origin of *Phalaenopsis* × *intermedia* Lindl. J Hortic Sci Biotechnol 81:691–699

Tsai CC, Sheue CR, Chen CH, Chou CH (2010b) Phylogenetics and biogeography of the *Phalaenopsis violacea* (Orchidaceae) species complex based on nuclear and plastid DNA. J Plant Biol 53:453–460

Tsai CC, Shih HC, Wang HV, Lin YS, Chang CH, Chiang YC, Chou CH (2015) RNA-seq SSRs of moth orchid and screening for molecular markers across genus *Phalaenopsis* (Orchidaceae). PLoS ONE 10 (11):e0141761

Tsai CC, Wu PY, Kuo CC, Huang MC, Yu SK, Hsu TW, Chiang TY, Chiang YC (2014) Analysis of microsatellites in the vulnerable orchid Gastrodia flavilabella: the development of microsatellite markers and cross-species amplification in *Gastrodia*. Bot Stud 55:72

Tsai CC, Chiang YC, Lin YS, Liu WL, Chou CH (2012) Plastid trnL intron polymorphisms among *Phalaenopsis* species used for identifying the plastid genome type of *Phalaenopsis* hybrids. Sci Hortic 142:84–91

Varshney RK, Graner A, Sorrells ME (2005) Genic microsatellite markers in plants: features and applications. Trends Biotechnol 23:48–55

Wang Z, Fang B, Chen J, Zhang X, Luo Z, Huang L (2010) De novo assembly and characterization of root transcriptome using Illumina paired-end sequencing and development of cSSR markers in sweet potato (*Ipomoea batatas*). BMC Genom 11:726

Young LG (2004) The study of microsatellite DNA identification in *Phalaenopsis*. Master Thesis, Institute of Forensic Science, Central Police University, Taoyuan Hsien, Taiwan. (Chinese with English abstract)

Yukawa T, Kita K, Handa T, Topik H (2005) Molecular phylogenetics of *Phalaenopsis* (Orchidaceae) and allied genera: reevaluation of generic concepts. Acta Phytotax Geobot 56:141–161

Zhang SM, Chen C, Chen FF, Wang T (2013) Analysis on genetic diversity of 16 *Phalaenopsis* cultivars using EST-SSR markers. J Plant Genet Resour 14:560–564. (Chinese with English abstract)

Zhu H, Senalik D, McCown BH, Zeldin EL, Speers J, Hyman J, Bassil N, Hummer K, Simon PW, Zalapa JE (2012) Mining and validation of pyrosequenced simple sequence repeats (SSRs) from American cranberry (*Vaccinium macrocarpon* Ait.). Theor Appl Genet 124(1):87–96

The manufacturer's authorised representative in the EU is Springer Nature Customer Service Centre GmbH, Europaplatz 3, 69115 Heidelberg, Germany. If you have any concerns regarding our products, please contact ProductSafety@springernature.com

Printed and bound by CPI Group (UK) Ltd, Croydon, CR0 4YY
30/09/2026
02225665-0020